Volume 12

SCOTTISH LOCOMOTIVE HISTORY

SCOTTISH LOCOMOTIVE HISTORY

1831–1923

CAMPBELL HIGHET

Routledge
Taylor & Francis Group

LONDON AND NEW YORK

First published in 1970 by George Allen & Unwin Ltd

This edition first published in 2022
by Routledge
2 Park Square, Milton Park, Abingdon, Oxon OX14 4RN

and by Routledge
605 Third Avenue, New York, NY 10158

Routledge is an imprint of the Taylor & Francis Group, an informa business

British Library Cataloguing in Publication Data
A catalogue record for this book is available from the British Library

ISBN: 978-1-03-206184-9 (Set)
ISBN: 978-1-00-321338-3 (Set) (ebk)
ISBN: 978-1-03-207771-0 (Volume 12) (hbk)
ISBN: 978-1-03-207773-4 (Volume 12) (pbk)
ISBN: 978-1-00-320871-6 (Volume 12) (ebk)

DOI: 10.1201/9781003208716

Publisher's Note
The publisher has gone to great lengths to ensure the quality of this reprint but points out that some imperfections in the original copies may be apparent.

Disclaimer
The publisher has made every effort to trace copyright holders and would welcome correspondence from those they have been unable to trace.

SCOTTISH LOCOMOTIVE HISTORY: 1831-1923

By Campbell Highet

London

GEORGE ALLEN & UNWIN LTD

RUSKIN HOUSE   MUSEUM STREET

FIRST PUBLISHED 1970

© *George Allen & Unwin Ltd 1970*

ISBN 0 04 625004 2

PRINTED IN GREAT BRITAIN
in 11 on 12 pt. Plantin
BY W. & J. MACKAY AND CO. LTD
CHATHAM, KENT

Acknowledgements

The author gratefully acknowledges the assistance rendered to him in gathering together, and ensuring the accuracy of, the data in this book. In particular he is indebted to E. T. Bryant, Borough Librarian, Widnes; the Technical Section of Liverpool City Library; Manchester Central Reference Library; and Robert Hogg and Staff of the then B.R.B. Historical Records, Edinburgh. Also the many friends who have helped, among whom are A. G. Dunbar, D. L. Smith, M. Smith and others. Illustrations are duly acknowledged but special mention is due to C. W. Black, City Librarian, Mitchell Library. Glasgow; the Director, Glasgow Museum of Transport; Dr R. L. Hills, Institute of Science & Technology, University of Manchester; and John Edgington of the Public Relations Office, Euston House, L.M.R. Not least my thanks are due to Rowland C. Bond, lately General Manager, British Railways Workshops, who has very kindly contributed a foreword.

CAMPBELL HIGHET

Wallasey and Derby, 1970

Foreword

It is almost exactly half a century since, in September 1920, Campbell Highet and I were taken by a member of Sir Henry Fowler's staff office at Derby Locomotive Works to the Brass Foundry, there to be introduced to the foreman of the shop to which we had both been allocated at the commencement of our apprenticeship. We were fortunate in our choice of Derby as the works at which to receive our practical training as locomotive engineers. The Midland Railway alone among the major British railway companies before the amalgamations of 1923, did not require the payment of a premium, commonly £150 per annum, a lot of money in those days, for the facilities which Derby offered. Moreover the training which we received combined practical work in the shops with attendance during working hours on two mornings each week at the local Technical College. There we were given a thorough grounding in the academic side of our chosen profession. The Midland was well ahead of the times in its approach to the training of its young engineers.

It is a far cry to those early years in Derby when the pride of the Midland locomotive stock were the three-cylinder compounds and the ten-coupled banker for the Lickey Incline. Paget's eight-cylinder 2-6-2 still lay under tarpaulin sheets in the Paint Shop hidden from the prying eyes of apprentices who were bold enough to explore the fascinating mysteries of a great locomotive works beyond the walls of their own shop. Our paths have diverged since those days, but we have both devoted the whole of our working lives to the service of the steam locomotive.

An efficient transport system is an essential prerequisite as much for the continued progress of developed industrial nations, as for opening up and exploiting virgin territory. Who can doubt that in this service to mankind the steam locomotive has played a predominant part?

To people like Campbell Highet and myself it is sad that our grandchildren will be able to learn something of the grandeur of a modern steam locomotive only by visits to the Railway Museum at York, the Science Museum at Kensington and other Transport Museums, of which there is one in Glasgow, where through the persistent efforts of many devotees of the steam locomotive, some famous examples are preserved. Not for them the unforgettable sounds, sights and smells compounded of hot oil and coal dust, of steam locomotives in service. Nor will they know the feel of riding on the footplate of a 'Duchess' Pacific, or even a humble Caley 0-6-0, the thrill of which was only enhanced the more one was able to indulge this harmless pastime. No other mechanical engineering development has been able to evoke so widespread and sustained an interest among young and old alike, inside and outside the railway service as the steam locomotive. It is important therefore that the whole absorbing history of steam motive power, to the perfecting of which Scotsmen, whether as railway engineers or locomotive builders, made a major contribution, should be faithfully recorded for posterity.

In his book, to which it is a privilege and pleasure for me to write this short foreword, Campbell Highet has done much more than describe the technical features of many of the classes of locomotives which did their work in Scotland. He has written about the men who directed affairs—great names like Stroudley, the Stirlings, Dugald Drummond and John F. McIntosh, to name but a few. In moving, as so many of them did, from one railway to another their influence extended over a very wide field, often far south of the Border. And with family connections involved, fathers and sons and brothers, there was continuity of technical development, and a corresponding family likeness in the external appearance of so many Scottish engines. We are told, too, something of the history of the railways of Scotland. We are reminded of the great races to the North over the West Coast and East Coast Routes, in which the Scottish partners played so prominent a part. It was on Scottish railways that the first 4-6-0 and the first four-cylinder locomotives ran in Great Britain—forerunners of most significant developments in later years.

One tends to forget the locomotive tests of earlier years in which locomotives, sometimes from another company, were matched against each other. The author describes many of the important tests conducted during the period of his review. But he has not told us of those days in 1924, in the early years of the

London, Midland & Scottish Railway, when a Caledonian 4-4-0, No. 124, was engaged in mortal combat with Midland Compounds and a 'Claughton' in comparative trials on express passenger trains between Leeds and Carlisle. No. 124 made a gallant effort and just about managed to keep time with 300 tons—but never before or since have I seen such a display of fireworks from any locomotive chimney.

I have been fortunate in knowing at first hand some of the more modern locomotives so well described in this book. Residence in Glasgow as an inspector at locomotive building works north of the Border, and nearly two years in charge at St Rollox, gave me opportunities for getting to know and evaluate the qualities of Scottish locomotives. Let me conclude with a word about the Royal Scots which I inspected during construction at the Queen's Park and Hyde Park Works of that great combine, the North British Locomotive Company. All that was best in British locomotive engineering and Scottish craftsmanship were combined in those fine locomotives, which though built after the years covered by Campbell Highet's book may yet stand as a fitting tribute to the long line of steam locomotives built in Scotland.

ROWLAND C. BOND

May 1970

Contents

FOREWORD   *page* 9

I   Historical Prologue   17

II   Initial Stages   35

III   The Trend-setters   50

IV   The Third Trend-setter   69

V   A New Star in the Firmament   88

VI   The Trends Develop   116

VII   James Manson Leaves His Mark   147

VIII   The Drummond Trend Widens   167

IX   John Farquharson McIntosh   187

X   The Introduction of Superheating   205

INDEX   235

Illustrations

1. Monkland & Kirkintilloch Railway, 1831 *facing page* 32
2. Glasgow & Garnkirk Railway, 0–4–0 *George Stephenson* *between pages* 32–3
3. Glasgow Paisley & Greenock Railway, 2–2–2 Caird & Co., 1843 32–3
4. Glasgow Paisley Kilmarnock & Ayr Railway, 2–2–2 *Lightning*, 1846 *facing page* 33
5. Caledonian Railway, 2–2–2 No. 15, Vulcan Foundry 1848—elevation 48
6. C.R., 2–2–2 No. 15—longitudinal section 48
7. C.R., 2–2–2 No. 15, Vulcan Foundry 1848—plan *between pages* 48–9
8. C.R., 2–2–2 No. 15—cross sections 48–9
9. Great North of Scotland Railway, Clark 2–4–0 No. 4, 1854 48–9
10. G.N.S.R., Cowan 4–4–0 No. 29, 1862 48–9
11. C.R., Conner 2–4–0, 158 class built Beyer Peacock, 1861 *facing page* 49
12. Edinburgh & Glasgow Railway, 2–2–2 No. 23, built Beyer Peacock, 1856 80
13. North British Railway, Hawthorn 0–6–0 No. 67A of 1850, as rebuilt by Wheatley, 1869 *between pages* 80–1
14. Highland Railway, Stroudley's largest snow plough attached to leading 2–4–0 of four 80–1
15. Scottish North Eastern Railway, 2–2–2 No. 461, Vulcan Foundry 1865 *facing page* 81
16. C.R., Conner Improved Goods, 2–4–0 No. 214, 1862 *facing page* 96
17. N.B.R., Wheatley 0–6–0 No. 363, 1869 *between pages* 96–7
18. N.B.R., Drummond 0–6–0T No. 96, 1875 96–7
19. William Cowan, G.N.S.R. *facing page* 97
20. Benjamin Conner, C.R. 97

21. David Jones, H.R.   128
22. Dugald Drummond, C.R.   128
23. Glasgow, Bothwell, Hamilton & Coatbridge Railway, 0–6–0T built Dübs & Co., 1877   *between pages 128–9*
24. H.R., 4–4–0 No. 7 for Dingwall & Skye line   *between pages 128–9*
25. C.R., 2–2–2T No. 1200 as rebuilt 1881   *facing page 129*
26. Glasgow & South Western Railway, Whitelegg rebuild of Smellie 'Wee Bogie', 1921   144
27. G. & S.W.R., Smellie 'Big Bogie' No. 77 with Bogie tender   *between pages 144–5*
28. H.R., Jones 4–6–0 No. 103   144–5
29. G.N.S.R., Rail Motor No. 29/28, 1905   *facing page 145*
30. G. & S.W.R., Rail Motor No. 1, 1905   176
31. G. & S.W.R., Rail Motor No. 3, 1905   *between pages 176–7*
32. James Manson with wife and friend   176–7
33. N.B.R., 4–4–2T, Yorkshire Engine Co., 1911   *facing page 177*
34. H.R., Smith 4–6–0 No. 70, 1915   192
35. G.N.S.R., Heywood 0–4–2T. built Manning Wardle 1915   *between pages 192–3*
36. G. & S.W.R., Whitelegg 'Baltic' tank No. 545, N.B.L. Co. 1922   192–3
37. G. & S.W.R., Whitelegg rebuild of Manson 4-cyl, 1922   *facing page 193*

Chapter I

HISTORICAL PROLOGUE

At the beginning of the nineteenth century the Industrial Revolution was under way in Scotland. The first stirrings of the vast changes which were to alter the lives and habits of the people are to be found in two events which took place in 1759. First the Carron Ironworks near Falkirk were founded by Samuel Garbett and John Roebuck, respectively an Englishman and a Scot; in this they were aided by William Cadell. Second, Glasgow Town Council obtained an Act of Parliament authorizing them to cleanse, enlarge and improve the channel of the River Clyde.

Scotland was rich in mineral deposits and in water power, but she lacked two essentials to successful development of her resources: craftsmanship and investment capital. Since the Union of Parliaments in 1707 there were ample trading opportunities but few good harbours, and such harbours as there were, through which to exploit these channels, were too far from the chief towns. In the half-century which followed the starting of the Carron Ironworks there were notable advances in agriculture and textiles as well as in the coal and iron industries.

In 1765 James Watt established the principle of the separate condenser and thus gave us the steam engine in a form suitable for general use. In England, Henry Cort evolved new methods of puddling and rolling wrought and malleable iron. About the same time the new cotton industry, which was partly mechanized, took firm root and mills were established at a number of places in Lanarkshire, Renfrewshire, and the Vale of Leven. Lanarkshire was the scene of another development in 1801, when David Mushet discovered the rich deposits of blackband ironstone. This ore, sometimes known as 'mushet-stone', had a high coal content, a factor which contributed to more economical smelting. The success of the Carron Ironworks gave rise to the establishment of other ironworks at Wilsontown, Omoa and the Clyde works (all in

Lanarkshire), and at Muirkirk and Lugar in Ayrshire. Other works followed in the early years of the nineteenth century.

The period also witnessed an improvement in road communications, due largely to the work of Thomas Telford on behalf of the Highland Commission for Roads and Bridges. Complementary to this was John Rennie's work on the construction of canals and better harbours, with which Telford also was associated. The year 1790 saw the Forth and Clyde Canal completed, a 35-mile waterway from Grangemouth on the Firth of Forth to Bowling on the River Clyde. There was a 3-mile branch to Port Dundas in north Glasgow. The same year the Monkland Canal, which joined it and ran eastwards for 12 miles to Airdrie, was opened. The Monkland's main function was the cheap transport of coal to Glasgow, and that of the Forth and Clyde was the transit of bulk grain in a westerly direction and of coal and machinery eastwards.

It was not until 1822 that the Union Canal, the remaining waterway in the central belt, was opened affording a connection between Falkirk and Edinburgh. The two Highland canals, the Crinan and the Caledonian, were opened in 1801 and 1822 respectively, and the Inverurie Canal in 1807. With the object of eliminating the difficult passage for ships from the inner estuary of the Clyde up the shallow river to Glasgow, a passage only possible by shallow-draught vessels, the project of a canal from Ardrossan to Glasgow was raised by the Earl of Eglinton and his supporters. The people of Glasgow were, however, intent on improving their river and gave little support to the plan.

Nevertheless, an Act was obtained and work commenced from the Glasgow end. Shortage of funds caused the termination of the canal at Johnstone, near Paisley, this portion being brought into use in 1811. At Ardrossan a considerable sum of money was spent in making improvements to the harbour and thus the matter rested for several years.

In 1812, Henry Bell's *Comet*, the first successful steamship to sail upon river waters and open seas, made its impact on the industrial scene and gave rise to a new industry as well as a new form of transport. The steamship proved a fresh stimulus to seaborne commerce and at the same time fostered the founding of many new firms which engaged in shipbuilding and marine engineering. Among the latter were a number of companies which later undertook the construction of steam locomotives and were the training grounds of many men destined to become well known in the field of locomotive engineering.

Despite the improvements in internal transport wrought by these activities, there was still a great need for better facilities for moving raw materials and manufactured products in greater bulk and more economically as the industries grew.

In many colliery areas, plateways had been established for the trucking of coal from the pits to the harbours on the coasts or the dock basins of the canals. Examples of these plateways were to be found in the Lothians, in Fife, in the Glasgow area and, perhaps best known of all, between Kilmarnock and Troon. The earliest known plate- or tramway was at Cockenzie and dealt with coal from the Tranent collieries. It had been established soon after the 1715 rebellion.

Near Dunfermline, Lord Elgin's collieries at Elgin and Wellwood were served by a plateway on which coal was conveyed to the harbour at Limekilns on the Firth of Forth. Similarly, further up the Forth at Alloa there was a plateway from the Alloa pits and, near Falkirk, the Carron Ironworks were connected with the pits of the Kinnaird colliery.

In 1812 the Duke of Portland established his plateway between Kilmarnock and Troon in Ayrshire, thereby ensuring that he had a good outlet through which the coal from his Kilmarnock pits could be exported from the harbour of Troon which he also developed.

It will be apparent from the foregoing that in Scotland railways in their primitive form began as a coalpit adjunct. Their immediate and subsequent development was in a subordinate role to the canals. The Monkland & Kirkintilloch Railway, authorized in 1824, was planned to transport coal from the collieries to the Forth and Clyde Canal and to Glasgow. The Ballochney Railway (1826) and the Slamannan (1835), an extension of the Ballochney which met the Union Canal, were similar but the Garnkirk & Glasgow (1826) was the first which openly set out to compete with canal transport and throughout its course it followed the line of the Monkland Canal. These railways opened up new coalfields and played an impressive part in the economic development of the area.

As early as 1825 there had been a proposal to link Edinburgh and Glasgow by a railway and the supporters of the project, impressed by the success of the Liverpool & Manchester Railway, revived the scheme in 1830 and were successful in obtaining authorization for a railway in 1838, and on February 16, 1842, the Edinburgh & Glasgow Railway was opened throughout its length.

Meanwhile, the railway system was being extended in the southwest. The proprietors of the Ardrossan Canal abandoned this scheme and turned their efforts towards a railway from Ardrossan to Johnstone, where it would connect with the completed portion of the canal. This, too, met with little success, only the 5½ miles from Ardrossan to Kilwinning being built.

The towns of Paisley and Johnstone were both important centres of the cotton industry and, to facilitate the transport of raw materials from the Clyde ports to the mills, and of the finished goods in the reverse direction, a railway to connect Glasgow, Paisley and Greenock was proposed. About the same time, though for a different reason, there was a proposal for a railway from Glasgow to Ayr. The main purpose of this line was to carry fresh farm and agricultural produce to the city area, as well as to afford to the town dwellers easy access to the coastal resorts. Thus was born the Glasgow, Paisley, Kilmarnock & Ayr Railway which, over its first seven miles, was to be a joint company with the Glasgow, Paisley & Greenock; these lines received their authorization in 1837 and opened in 1840 and 1841 respectively. The G.P & G. subsequently became part of the Caledonian Railway which was to become the dominant power in the western part of the country.

The need for rail communications with industrial England was apparent in the 1830s and the first steps were taken towards the formation of a company to construct a line to connect Glasgow and Edinburgh with the English railway system, which at that date was still ill-defined and far from assuming the proportions ultimately reached. The Grand Junction proprietors envisaged the extension of their line towards the border by means of the Lancaster & Preston and Lancaster & Carlisle Railways, and at their behest Joseph Locke surveyed a route from Carlisle to Glasgow following the course of the River Annan to Beattock and thence via the Clyde valley, with a branch from the vicinity of Symington to Edinburgh. Not altogether liking the severe gradients involved in the Scottish uplands around Beattock, Locke transferred his attentions to the more easily graded route up the valley of the Nith. This course was the one favoured by the sponsors of a rival company (the Glasgow, Dumfries & Carlisle Railway) which had been promoted by the Glasgow & Ayr faction. Then the government took a hand by setting up a Royal Commission to examine the question of communications between England and Ireland and England and Scotland. The terms of reference included a specific instruction to give due consideration to transport facilities between Glasgow and the manufacturing districts of Lancashire. A long and strenuous fight ensued between the supporters of both the Annandale and Nithsdale routes, nor were the contending parties appeased when the Commissioners reported in favour of the former, but the Annandale party were jubilant when on July 31, 1845, the Caledonian Railway Bill received the Royal Assent. Farsighted speculators had their eyes fixed on Aberdeen as their northern goal and it was reached by a series of connecting railways which became, in their turn, parts of the Caledonian 'empire'. From Garriongill Junction, a few miles south of Motherwell to Gartsherrie, the old Wishaw & Coltness line of 1829 was followed, thence to Greenhill was to be a part of the new Caledonian line, and forward from Greenhill to Perth the Scottish Central Railway provided the route. At Perth, a connection was made with the Dundee & Perth and the Scottish Midland Junction Railways, then by the Arbroath & Forfar the line was continued to Guthrie where another end-on junction with the Aberdeen Railway enabled a continuous line of railway from London (Euston) to be achieved by April 1, 1850.

Glasgow itself was entered first of all at Glebe Street station, the terminus of the Garnkirk & Glasgow, but this was replaced by Buchanan Street station on November 1, 1849. On the south side of the Clyde there was another station, called simply 'South Side', which handled the local traffic to Hamilton, Barrhead, etc., leaving Buchanan Street to deal with the Edinburgh and English traffic. The coast trains were handled at Bridge Street station, the joint station of the G.P. & G. and G.P.K. & A. Railways. Branch lines were made wherever it was thought that traffic might offer and generally with an ulterior motive as, for example, the Dumfries Lochmaben & Lockerbie of 1860 which, at Dumfries in G. & S.W.R.

territory, gave the Caledonian a direct connection with Ireland by the Portpatrick Railway. Running powers over the G. & S.W.R. between Dumfries and Castle Douglas were obtained in 1864.

The Caledonian 'empire' grew with the absorption of a large number of smaller companies, chief among which were the Dundee, Perth & Aberdeen Junction (itself an amalgam of the Dundee & Perth and Dundee & Newtyle); the D.P. & A.J. was taken over by the Scottish Central in 1863 and this company was, in turn, absorbed by the Caledonian in 1865. The Scottish North Eastern, made up of the Scottish Midland Junction and Aberdeen Railways, was brought into the larger company in 1866. The last major project was the Callander & Oban Railway which, worked and maintained by the Caledonian, was nevertheless a separate entity until taken into the L.M.S. in 1923. Endeavours to establish a West of Scotland Junction station in the centre of Glasgow as early as 1846 met with no success, and not until the 1870s was the opposition of the Admiralty to the bridging of the Clyde finally overcome and the new terminal, Central, opened in 1879. Growth of traffic was such that enlargement was necessary in 1890 and again in 1905.

During the early years of the G.P.K. & A. the board had little time for matters other than those of immediate concern with their own undertaking. Nevertheless, some of the directors were very active on the provisional committee of the Glasgow, Dumfries & Carlisle which was to be a continuation of the Ayrshire line into England. The controversial issue of the route to be followed by the rival Caledonian faction having been settled in that company's favour acted as a spur to the G.D. & C. supporters who were successful in obtaining an Act on August 13, 1846, but only after modification of their proposals. The result of this was that the Nithsdale route to Carlisle joined the Caledonian, or Annandale route, at Gretna and entered Carlisle by the exercise of running powers over the last nine miles. The Ayrshire and Dumfries companies were amalgamated to form the Glasgow & South Western Railway from October 28, 1850, when the lines were opened throughout.

The Ayrshire line pushed south to Maybole and Girvan by 1856 and 1860 respectively. From Dumfries, the Castle Douglas & Dumfries Railway of 1859 and the Portpatrick Railway of 1861 went to the formation of the large delta-form that the G. & S.W.R.

presented on the map, but the triangle was not completed until the Girvan & Portpatrick Junction was opened in 1876. Absorption of the smaller lines by the parent company took place at several times; the Maybole & Girvan, Castle Douglas & Dumfries and Kirkcudbright Railways in the south, and the Bridge of Weir and Greenock & Ayrshire in the north all coming into the fold in 1865, followed by the Ayr & Maybole in 1871. The two northern lines, the Bridge of Weir and Greenock & Ayrshire, gave the G. & S.W.R. access to the Clyde steamers at Greenock and sparked off a vicious rivalry between that company and the 'Auld Enemy', the Caledonian, who operated from Gourock and Wemyss Bay. To meet this competition, the Sou' West made a line from Kilwinning to Ardrossan, West Kilbride, Fairlie and Largs, the latter point being reached in June 1885.

The direct route from Glasgow to Kilmarnock via Barrhead was opened in 1873 and was a joint concern with the Caledonian. By it, the distance to Carlisle was shortened by ten miles, a valuable asset when the through expresses between London (St Pancras) and Glasgow (St Enoch) started running in 1876. These trains used the new and handsome terminus in St Enoch Square, Glasgow. This station was almost doubled in size at the turn of the century, so greatly had traffic developed.

In conjunction with the North British Railway, the G. & S.W.R. was a partner in the City of Glasgow Union Railway, which made a valuable contribution to the city's suburban transport problem until the growth of street tramcar services gave rise to the partitioning of the City Union between the owning companies and, later, to the withdrawal of passenger services.

An interesting event in the company's history was the filling in of the Johnstone Canal and its reconstruction as a railway. This work was completed in 1885 and gave the G. & S.W.R. a useful additional line between Glasgow and Elderslie, west of Paisley, for the very heavy coast traffic, further aided by the North Johnstone line to Brownhill Junction near Dalry, opened in 1905.

The North British was the largest of the five main Scottish companies and, like its contemporaries, grew from a multiplicity of smaller lines. In its final form it comprised some fifty of these lesser undertakings, brought into the parent organization by purchase or amalgamation. The first real attempt to connect the Scottish capital with places south of the border was due to the expansionist efforts of George Hudson who saw extension to Edinburgh as the logical sequel to the making of his Newcastle & Berwick Railway.

The Railway Commissioners, Colonel Frederic Smith and Professor Peter Barlow, had pronounced in favour of the Annandale route on the west side of the country and it appeared as though the eastern side might be left without a rail connection to Scotland, since the view was held by many people that only one such line was necessary. Scottish opinion was contrary to this, whatever the government's attitude might be, and proposals for a line from Edinburgh to Berwick were launched in 1843. The authorization of the line, to be known as the North British Railway, was obtained on July 19, 1844, and it was opened on June 18, 1846. Rail communication between London and Edinburgh, via Rugby, Derby, York and Newcastle, was thus complete save for the water gaps at Newcastle and Berwick, and these were bridged in 1850. Six weeks after the opening of the N.B. main line, the Edinburgh & Glasgow extended its line from Haymarket into the North Bridge station in the capital and brought Glasgow into the eastern line of communication. By the building of branches and connecting lines in the central industrial belt, the E. & G. and N.B. companies were able to tap the mineral traffic of the country.

In 1847, another company closely associated with the N.B. began operations giving access to the Fife coalfields and paving the way towards Perth and Dundee. This was the Edinburgh & Northern, which ran from Burntisland on the north shore of the Firth of Forth to Ferry-Port-on-Craig, later renamed Tayport. From here, a ferry gave a connection to the Dundee side of the Tay at Broughty Ferry. The crossing of the Forth was also effected by a ferry, which operated between Granton and Burntisland and, so far as freight trains were concerned, this was a train ferry, the first of its kind in the world, designed by Thomas Bouch, the E. & N. engineer.

The Edinburgh & Northern was reincorporated in 1851 as the Edinburgh, Perth & Dundee, and this company amalgamated with the N.B. in 1862. Fife became criss-crossed by a number of lines including the Fife & Kinross, the Kinross-shire, Leven & East of

Fife, and the St Andrews Railways, all of which passed into the North British orbit in the course of time.

Southwards from Edinburgh what became known as the 'Waverley' route had its beginnings in the Edinburgh & Hawick Railway, opened by stages in 1847–9. Carlisle was reached in 1862 by means of the Border Union Railway and in the same year the Border Counties line was completed, providing a connection with the Newcastle & Carlisle section of the North Eastern Railway. A number of other lines in the border country, such as the Wansbeck Valley, the Berwickshire, and the Northumberland Central, opened in the 1860s, and branches to the important towns, e.g. Selkirk, Jedburgh, Roxburgh and Peebles, all contributed to the enlargement of N.B. influence, which was dominant, in the area. Around Carlisle there were the Port Carlisle and the Silloth Bay companies dating from 1854 and 1856. From the latter there was a sea connection with Ireland operated from the docks at Silloth and both companies came into the North British fold in 1880.

The bridging of the Tay was accomplished in 1878, though this triumph was short-lived. Some of the spans collapsed during the great storm of December 28, 1879. The present bridge was opened in 1887, leaving only one water gap to be bridged on the route towards Aberdeen. This last achievement, one of the wonders of the world, was the immense Forth Bridge. The culmination of a quarter of a century of discussion, negotiation and planning, it was opened on March 5, 1890.

The continuation of the northern drive to Aberdeen was obtained over the lines in Angus from Dundee via the Dundee & Arbroath, jointly owned by the N.B. and Caledonian, to a head-on junction with the North British, Arbroath & Montrose. Despite efforts to reach the Granite City over its own lines, the N.B.R. only reached its objective by exercising running powers over the 38 miles of Caledonian line from Kinnaber Junction, a circumstance which was to cause the N.B. some embarrassment during the races to the north in 1895. In both the Edinburgh and Glasgow districts a number of branch and suburban lines were made to meet the needs of the rapidly growing residential areas around the cities. One such line in particular was destined to become part of one of the country's most scenic routes. In 1862, the E. & G.R.

absorbed the Glasgow, Dumbarton & Helensburgh, including the once isolated Caledonian & Dunbartonshire Junction Railway of 1846, and in 1894 the Helensburgh line became part of the West Highland Railway, spectacularly engineered up Glen Falloch and across the Moor of Rannoch to Fort William on Loch Linnhe. The extension to Mallaig on the extreme west coast was carried out in 1901.

There were also several small lines in Stirlingshire and Dunbartonshire; one of these, the Forth & Clyde Junction Railway was opened in May 1856 and retained its independence until it was taken into the L.N.E.R. at the 1923 grouping, although leased to the N.B.R. from 1871. Like the Caledonian, the N.B.R. had a line through the centre of Glasgow, mostly underground and made on the 'cut-and-cover' principle. This was the Glasgow City & District line, opened in 1886, and which linked the eastern and western parts of the city.

In association with the G. & S.W.R. the N.B. owned part of the City of Glasgow Union Railway which afforded a rail connection north and south across the city, from Springburn to Shields Road and the Harbour General Terminus. The suburbs of Edinburgh were likewise well catered for by the Southside & Suburban Railway of 1884 and several other branches.

Like the Caledonian and G. & S.W. the N.B. was an active participant in the highly competitive steamer traffic on the Clyde, operating from a pier at Craigendoran, near Helensburgh, and the company had extensive docks at Granton, Burntisland, Methil and Alloa on the Forth, and at Stobcross in Glasgow. Waverley station, Edinburgh, grew from the original North Bridge station which housed the N.B. and E. & G. companies and which was enlarged several times. It was the principal station on the line and was the largest railway station outside London. In Glasgow, Queen Street, the original terminus of the E. & G.R., likewise underwent enlargement. It was reached by the formidable Cowlairs incline, up which trains were hauled by a stationary engine operating a cable, until 1909, save for a brief spell from October 1844 until March 1847 when special banking engines were in use. At Carlisle, the N.B.R. was able to use the Citadel station by an agreement of 1860 and a similar measure in 1864 related to Perth (general) station.

The proposals for a railway to connect Inverness and Perth by a line crossing the Grampians met with little success in 1845 and caused the promoters to abandon the idea of a direct line in favour of a more roundabout route via Aberdeen. Sixteen years were to elapse before the Inverness & Perth Junction Railway was authorized to make a line from the highland capital to connect with the Perth & Dunkeld Railway which had been opened in 1856. The I. & P.J.R. was opened throughout on September 9, 1863, Perth being reached over the absorbed P. & D.R. and the last seven miles from Stanley Junction being covered by running powers over the Scottish Central which owned this length. In the interim period the Inverness & Nairn Railway had been formed and opened in November 1855 and by 1858 had been extended to Keith and become the Inverness & Aberdeen Junction Railway. The Great North of Scotland Railway having reached Keith in 1856, rivalry between these companies had the result that neither reached its objective save by running over the other's lines. Amalgamation of the I. & A.J. and I. & P.J. Railways was effected in 1863, when the title 'Highland Railway' was adopted.

Northwards and westwards from Inverness there were several small companies, all of which were brought into the Highland net in 1884. The Ross-shire reached Invergordon in 1863 and Bonar Bridge the following year; the Sutherland Railway carried the line on to Golspie by 1868. Thence to Helmsdale the $17\frac{1}{2}$-mile stretch was constructed by the Duke of Sutherland and was opened in 1871. Continuation of the line to Wick and Thurso followed in 1874 when the Caithness Railway was opened, thus completing the railway communication with London, 722 miles to the south. From Dingwall, the Dingwall & Skye was opened in 1870 as far as Strome Ferry which remained the western terminus until the extension to Kyle of Lochalsh was completed on November 2, 1897.

In 1898 the direct line from Inverness to Aviemore, via Carr Bridge, was opened thereby shortening the distance to Perth by $25\frac{3}{4}$ miles. Nevertheless, this line introduced a third high summit at Slochd, 1,315 ft. above sea level, a formidable obstacle to southbound trains starting 'cold' from Inverness. The other summits were at Dava, 1,052 ft., on the old Forres–Aviemore line, and at Druimauchdair, 1,484 ft., on the line thence to Perth.

The original conception of the Great North of Scotland Railway was one of the fruits of the 'railway mania'. This somewhat ambitious main line was to reach from Aberdeen to Inverness via Huntly, Keith, Elgin and Nairn with branches to serve some of the towns on the Moray Firth. After the failure of the Bill put up by the Inverness party for a line over the mountains to Perth, the G.N.S.R. supporters obtained an Act for their line on June 26, 1846.

Construction of the G.N.S. began in 1852 after delays due to financial troubles, and the line was opened as far as Huntly in September 1854. The Aberdeen terminus was at Kittybrewster. From there a branch to the Waterloo Quay in Aberdeen harbour was opened in 1855. The western extension from Huntly to Keith was opened on October 10, 1856, completing the connection with the I. & A.J.R. to Inverness when that line reached Keith two years later. The Denburn Valley line from Kittybrewster to the Scottish North Eastern, near the Guild Street station in Aberdeen, was opened in 1867 and a new joint station was built to serve the companies running into the city.

Between 1851 and 1861 various branches were added. From Dufftown, the Strathspey Railway was opened in sections to Boat of Garten between 1863 and 1866; here physical connection was made with the Highland. All these subsidiary undertakings were absorbed by the G.N.S.R. by an Act of August 1, 1866. In 1867 the Banffshire Railway was brought into the fold. This line had started as the Banff, Portsoy & Strathisla Railway of 1859, serving part of the coastal strip along the Moray Firth.

Elgin was reached in 1863 by taking over the working of the Morayshire Railway, of which the first part constructed (from Elgin to Lossiemouth) opened on August 10, 1852 and was the oldest railway in the north of Scotland. In 1880 the Morayshire was amalgamated with the Great North.

As early as 1845 proposals had been made for a railway to serve the valley of the River Dee. A line from Ferryhill Junction at Aberdeen to Aboyne was authorized on July 16, 1846. By September 1853 it was opened as far as Banchory, after a series of financial difficulties, and Aboyne was reached in December 1859. The extension to Ballater was carried out by 1866. Ballater became the railhead for royal journeys, being the nearest point the railway reached to the Queen's estates around Braemar. Other small branches were added to the system from time to time, the last such being the Fraserburgh & St Combs Light Railway, authorized on September 8, 1899, under the Light Railways Act of 1896 and opened on July 1, 1903. As this line was unfenced it had the distinction of being one of the only two lines in Britain on which the locomotives carried cowcatchers. The other line was also in Scotland, but in the south—the Wanlockhead branch of the Caledonian.

Between 1880 and 1890 many improvements were carried out for the more efficient conduct of the company's business. Among these improvements were such items as the doubling of the main line and the widening of various parts of the system. Another development during this period was the automatic tablet-catching apparatus invented by James Manson, the locomotive superintendent. By its use, tablets for single lines of railway could be exchanged at speed without the exposure to injury that the enginemen and signalmen would otherwise be liable to suffer.

Of the five main railways of the Northern Kingdom, the G.N.S. was the smallest and most parochial. It served a locality almost wholly devoid of manufacturing industries and depended for its livelihood on the transport of agricultural produce and fish. Many of the small harbours on the northern coastline were fishing harbours and Aberdeen, besides being the principal fishing port, had a lively trade with the islands of the far north and with Leith and other southern ports. As the city grew, so its suburban traffic grew. With its Highland neighbour there was little exchange of traffic; nevertheless, despite the localized nature of the company, the G.N.S. earned a reputation for brisk business during its later years, in complete contrast to the manner in which it was viewed during its early days.

The development of the steam locomotive in Scotland, as has been the case throughout the whole history of railways, has been due to such factors as the exploitation of the mineral deposits of one area, the establishment of new industries in another, the growth of an urban or rural locality until it required its own rail services, the increasing weight and speed of trains, and so on.

The group of lines in the central part of the country, which become the Monkland Railways, together with their neighbours the Coltness and Clydesdale Junction portions of the Caledonian, played an important part in developing the coal and iron industries. For the haulage of these minerals, the N.B., Caledonian, and G. & S.W. Railways all had various locomotives designed for their services. As time went on, more powerful, and usually larger and heavier, locomotives became necessary. It was the same in the case of the Anglo-Scottish traffic, when the main trunk routes were opened. The little 2–2–2 locomotives of various types used on the passenger trains, and the 0–4–2 types on goods and mineral trains, were required to face banks on the main lines which taxed them severely, and probably nowhere worse than on the Highland main line over the Grampians. It was David Jones who was to revolutionize the motive power of this company, first by the introduction of his 4–4–0s in 1874 and then, in 1894, by his famous 'Big Goods' class, the first British 4–6–0s, which distinguished themselves on passenger as well as goods services. The neighbouring G.N.S.R., having found the 4–4–0 type so eminently suitable to their needs in Cowan's day, never departed from this wheel arrangement for their main line locomotives, no matter whether for express passenger, goods or fish trains.

All three southern companies had heavy commitments with mineral traffic. Coal from Ayrshire found an outlet from the ports on the Clyde estuary at Ardrossan, Irvine, Troon and Ayr over the G. & S.W.R. metals, whilst the Caledonian conveyed its traffic from the Ayrshire and Lanarkshire collieries to Grangemouth in the east. In order to obtain and develop this port on the Firth of Forth, the Caledonian had to purchase the Forth & Clyde Canal, which it did in 1867, thus gaining an important outlet on that side of the country—much to the chagrin of the N.B.R. whose main port at that time was Burntisland. It was not until 1887 that the N.B.R. acquired the harbour at Methil, in Fife, which was enlarged in 1897 and again in 1907. When the Midland Railway completed the Settle & Carlisle line in 1876 and commenced through services to Glasgow and Edinburgh over the G. & S.W. or N.B. 'Waverley' routes there was an upsurge of locomotive development.

Almost concurrent with these events was the realization that more attention to brakes was necessary. The Government had been hammering at the companies for years through the medium of the Board of Trade Inspecting Officers, only to be met with a kind of passive resistance on the part of many railways. However, the now famous Newark trials staged in 1875 by the Midland Railway set the stage for future development. At this time Dugald Drummond was locomotive superintendent of the N.B.R. and he arranged his own series of trials after the Newark event was over, and began fitting N.B.R. locomotives and vehicles with the Westinghouse brake. Meanwhile, the Caledonian had been experimenting with the Steel-McInnes air brake, and not until Drummond left the N.B. and came to the Caledonian in 1882 was any concerted effort made to adopt the Westinghouse system on this line.

On the G. & S.W.R., Smellie was fitting the Westinghouse brake as standard until such time as the Midland, having run into trouble with the Westinghouse Brake Company, changed over to the vacuum brake and the G. & S.W. followed suit, a not altogether surprising happening since at that time both railways were under the chairmanship of Matthew Thompson. The year 1880 saw the opening of the Callander & Oban line for which the Caledonian produced various designs of locomotive, each larger than its predecessors, in order to meet the growing traffic, mainly tourist, on this severely graded line.

The only Scottish company directly involved in train working during the races to Edinburgh in 1888 was the Caledonian. The results of the skirmish between the east and west coast routes were far reaching, and paved the way for many improvements in locomotive design. When, in 1890, the Forth bridge was opened and a new through route to Aberdeen was presented to the travelling public, a new vista of speedy rail travel began. As the largest city in north-east Scotland Aberdeen became the goal and the seal was set to this by the races of 95. The behaviour of the locomotives used on the northern side of the border was typical of the products of the man responsible for their design; for it was Dugald Drummond who brought out the basic design of both the N.B.R. and the Caledonian 4–4–0s which covered themselves with glory.

A man who learned much from these events and put the knowledge to good use was John Farquharson McIntosh, who introduced the famous 'Dunalastair' class on the Caledonian in 1896. In these engines he showed how well the lessons of the 1895 races had been absorbed, by giving the locomotive a boiler larger and more powerful than had been previously known. McIntosh's

31

successor, Pickersgill, carried on the good work and the effects spread also to other companies.

Heavier trains on the east side of the country resulted in the increasing size of the popular 4–4–0 and led to the large 4–4–2 designed by Reid in 1906. Apart from the G.N.S.R., the N.B.R. was the only Scottish company which never built a 4–6–0, the allegedly greater flexibility of the 4–4–2 being given as the reason for the choice of this type. Again, it may be noted that the Caledonian was the only Scottish company to produce an eight-coupled freight engine. McIntosh designed his 0–8–0 of 1901 specially for use with the 30-ton bogie open mineral wagons built for the coal and iron traffic. Both wagons and locomotives were fitted with Westinghouse brakes and possibly provide the first instance in this country of continuously braked heavy freight trains.

The North British had a traffic problem somewhat similar to that of the Caledonian's Callander & Oban line in its West Highland Railway. Here again, the severity of gradients and curvature gave rise to special designs. Much the same story applies to the trains run in connection with the Clyde steamship sailings. Especially so was this the case of the Caledonian and the Sou' West. On both companies, successive superintendents produced their own ideas of the panaceas for all the ills that beset them. A good example of this is to be found in the various G. & S.W. classes specially designed for the Greenock road by James Stirling, Smellie and Manson.

The cities were adequately furnished with good services to the suburban areas, though in the case of Glasgow particularly, the ever expanding and excellent tramway services afforded by the enlightened corporation severely affected the three main companies concerned, the N.B., G. & S.W. and Caledonian Railways. The spread of the tramways resulted in the withdrawal of many of their suburban services. Instead, the companies concentrated on the longer distance commuter traffic, e.g. to the coast towns, and gave services which had no parallel outside the London area. In this connection, the intensity of the G. & S.W.R. services to the Ayrshire coastal resorts has been likened to those of the L.T. & S.R. in the south, and the provision of power units to operate them culminated in Robert Whitelegg's massive *Baltic* tank engines of 1921.

1. Monkland & Kirkintilloch Railway, first Locomotive built in Glasgow, 1831.
    Glasgow Museum of Transport.

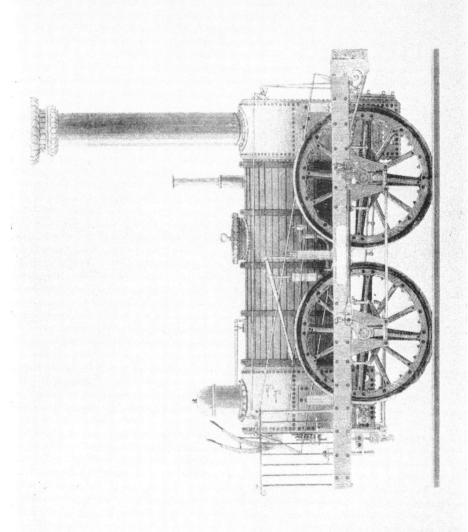

2. Glasgow & Garnkirk Railway, 0–4–0 'George Stephenson', built by Robert Stephenson & Co., 1831. Works No. 27

3. Glasgow, Paisley & Greenock Railway, 2–2–2 by Caird & Co., between 1843 and 1856. *Glasgow Museum of Transport.*

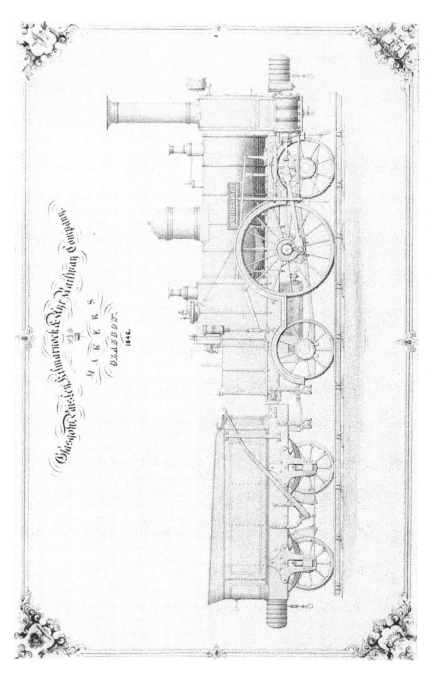

4. Glasgow, Paisley, Kilmarnock & Ayr Railway, 2–2–2 built by Cook Street Works of the Coy., No. 28 'Lightning', 1846. *Glasgow Museum of Transport.*

Thus each phase of railway operation, and each era through which the companies passed, produced their separate problems, each one a challenge to the locomotive superintendents of the time. The story is long and complex, and one in which the railway companies had close connections with the many firms of locomotive manufacturers in Scotland and in England.

While the railway industry was yet young it was the practice for the companies to purchase their motive power from one or more of the private engine building firms, relying on each firm's knowledge and experience, however circumscribed, of what was best suited to their needs. Then as they grew in experience and confidence themselves the companies began to dictate their requirements to the builders in greater detail. However, other factors began to assert themselves. The opening of the Great North of Scotland Railway was marred by the dilatoriness of Hawthorns of Leith in delivering the first engines for this line in 1854. Again in 1863, the Highland Railway minuted an explanation of the delay in the delivery of locomotives for the Inverness & Perth Junction Railway owing to the shortage of skilled staff at the makers' works. As a result the initial locomotive work of this line was performed by the Scottish Central.

After an essay in building some bank engines in its own Cowlairs shops the Edinburgh & Glasgow reverted to the practice of purchasing from outside firms until Steel Brown changed the pattern of things, after which both practices obtained. The same can be said of the other companies. Until Kilmarnock works replaced the old Cook Street establishment, the G. & S.W.R. bought from outside firms as did the Caledonian at St Rollox, save for a few engines built at Greenock.

Another factor which undoubtedly affected the issue was the extent to which money was, or was not, available. Throughout the history of railways the locomotive department has been considered a spending department, and as such has had to suffer grievously from time to time from tightly drawn purse strings. This pecuniary distress resulted in locomotive superintendents having to make use of any recoverable and repairable parts of old, and often obsolete engines and from them constructing new and often much improved machines.

That the construction of their own locomotives by the railway companies in their own shops was a bone of contention with the manufacturing firms was clearly brought out by James Reid, Chairman of Neilson & Co., when he gave his presidential address to the Institute of Engineers & Shipbuilders in Scotland in Glasgow on October 24, 1882. Amongst the beliefs he held he maintained that the railways were common carriers and should confine their endeavours to following that course, obtaining their locomotives from the firms specializing in that branch of engineering in which they had the necessary expertise. Further, he held that the locomotive engineers of the railways had enough to occupy them in the maintenance and repair of their engines and in the efficient organization and working of their departments.

James Reid's views do not appear to have influenced the railwaymen amongst his audience for the companies continued to build their own motive power units, though at the same time it must be stated that a high proportion of their needs was met by contract firms and most amicable relations have been maintained between all parties with free interchange of information and ideas despite the criticisms of Reid.

How the railwaymen and the specialist manufacturers met the challenges of each succeeding age and coped with the difficulties confronting them, and produced studs of locomotives which stood high in the annals of the industry, often leading the way for others to follow, is told in the following pages.

Chapter II

INITIAL STAGES

The first successful use of steam locomotives in Scotland was on the Monkland & Kirkintilloch Railway in the year 1831 and was the result of something in the nature of a race between this company and the Garnkirk & Glasgow Railway. The latter was known to have ordered some of Stephenson's 'Planet' type of locomotives for the opening of its line. Accordingly the M. & K. directors instructed their locomotive superintendent, Isaac Dodds, to produce designs for two locomotives capable of hauling 60 tons gross weight at a speed of 5 m.p.h. Dodds based his design on the Killingworth type and an order was placed with Murdoch & Aitken, Hill Street, Glasgow, who thus achieved fame as having constructed the first steam locomotives to be built in Scotland.

The builders delivered the first engine on May 10, 1831, the second one following on September 10th the same year. The boilers of these engines were 4 ft. 6 in. diameter and contained 62 tubes $1\frac{1}{2}$ in. diameter and 5 ft. long, while the working pressure was 50 lb. per sq. in. and the firegrate 4 sq. ft. area. The cylinders were $10\frac{1}{2} \times 24$ in. and the coupled wheels, of which there were four, were 3 ft. 9 in. diameter and were provided with 1 in. side play. The coupling rods had ball and socket joints and the pistons were fitted with metallic packing, one of the earliest examples of its use in the locomotive field. The packing consisted of two iron rings, in three segments, a wedge piece between each pair of segments being pressed outwards by means of a spiral spring. The valves were operated by tappet gearing moved by eccentrics. Although the Killingworth type of locomotive was now being looked upon as having been surpassed by the 'Planet' type designed by Stephenson, these Murdoch & Aitken engines for the M. & K. were considered highly efficient machines by the directors of the company. They bore the names *Monkland* and *Kirkintilloch* until later numbered 1 and 2.

When the Garnkirk & Glasgow Railway was officially opened later in 1831 two of Robert Stephenson & Co's 'Planet' type engines were employed. These were *St Rollox* and *George Stephenson*, maker's numbers 36 and 37. The former was a 2–2–0 and the latter an 0–4–0. *St Rollox* had two cylinders 10 × 14 in. and driving wheels 4 ft. 6 in. diameter. The boiler of this engine was 3 ft. diameter and 6 ft. in length and had a heating surface of 298·5 sq. ft. The pressure was again 50 lb. per sq. in. Dendy Marshall numbers this locomotive 70 in the first 100 built. The cost of this engine was stated to have been £780 and, after five years service on the G. & G.R., it was sold to the Paisley & Renfrew Railway for £350. When the locomotives of this railway were sold for scrap in 1848, *St Rollox* realized only £13.

The years 1832 to 1836 saw further additions to the locomotive stocks of both railways. Johnston & McNab supplied an engine of the 0–4–0 'Planet' type named *Glasgow* in 1832. This was followed shortly after by *Garnkirk*, another 'Planet', made by Murdoch & Aitken who also built a six-coupled engine *Gartgill* the following year. Johnston & McNab were the makers of the third and fourth locomotives which were followed in 1836 by *Jenny*, another 2–2–0 'Planet' shortly before the firm moved to Paisley. A small firm known as the St Rollox Foundry Company built an 0–4–0 for the same line in 1835; this bore the name *Frew*. Besides being the first six-coupled locomotive built in Scotland, *Gartgill* had a peculiar drive. The cylinders were mounted high up at the foot-plate end and inclined upwards. From the crosshead a connecting rod worked a crank which drove a second connecting rod coupled to a crank on the middle axle. The cylinders were $12\frac{1}{2} \times 21$ in., the wheels 3 ft. 9 in. diameter and the wheelbase was 10 ft. The boiler contained 90 tubes $1\frac{3}{4}$ in. diameter. *Frew* was of the same general design as *George Stephenson*, but *Jenny* was similar to *St Rollox*. It had only 85 tubes giving a heating surface of 294 sq. ft., and carried the customary 50 lb. per sq. in. pressure.

It is unfortunate that no details have survived of the first loco-motive to be built in a railway company's workshops in Scotland. The subject of this distinction was named *Victoria* and emerged from the M. & K. works at Kipps, near Airdrie, in 1837. Seemingly all that is known of it is that it was used for experimental work towing canal boats on the M. & K. Canal. Immediately, a fresh problem was posed: how to overcome the scouring action on the canal banks due to the wash of the fast moving boats and this brought the experiment to a sudden end.

That the railways were proving successful investments is demonstrated by the fact that it has been recorded that whereas the conveyance of coal, the main source of revenue, from the colliery areas around Airdrie and Coatbridge had been costing up to 3s 6d per ton for cartage to the canal and forward to Glasgow by boat, it could be taken direct to the city by rail for only 1s 3d per ton.

In 1816 or 1817 a Killingworth type locomotive was tried on the Kilmarnock & Troon line. An account of the proceedings has been left in his book 'Retrospect of an Artist's Life' by John Kelso Hunter, the Kilmarnock shoemaker-artist, who described what he recollected seeing when he was a boy some fifty years earlier. The lack of success of the Duke of Portland's enterprise in having one of George Stephenson's engines on his railway, was said to have been due to the manner in which the engine thumped the track causing serious damage by the breaking of the tram plates.

While these developments were taking place in the west, the Dundee & Newtyle Railway had obtained their first two locomotives from J. & C. Carmichael in 1835. The basic design employed was that of Richard Roberts of Sharp, Roberts & Co. whose 'Experiment' type of locomotive had been built for certain English and Irish companies. The D. & N. engines were *Earl of Airlie* and *Lord Wharncliffe*. A third engine is attributed to Stirling & Co's East Foundry, Dundee, and was of the 0–4–2 type and named *Trotter*. The design used vertical cylinders, but whereas on the Killingworth types the cylinders were usually mounted in the boiler, here they were outside the frames at a point opposite the joint between the boiler and smokebox. Eccentrics for the valve gear were not used, a derived form of motion being employed instead; a fulcrum pin on a bell crank on one side providing the motion of the opposite side valve. The valves were of a tubular form in wrought iron and unequal expansion of the inner and outer valve tubes gave rise to troubles which resulted in the abandonment of the idea. In later years Paget on the Midland and Bulleid on the Southern attempted to use sleeve valves but again the results were unsatisfactory.

In April 1836 Robert Stephenson & Co. built a fourth loco-motive for the D. & N.R. This was *John Bull*, maker's number 137, and was generally similar to its predecessors. After the D. & N. was altered to standard from 4 ft. 6 in. gauge, numbers 1, 2 and 4 were converted and taken over by the Dundee, Perth & Aberdeen Junction Railway in 1848 and were later used as pumping engines.

. The Paisley & Renfrew Railway was opened in 1837 and again the gauge employed was 4 ft. 6 in. Locomotives for this line were three in number. One was *St Rollox*, mentioned above, purchased from the G. & G.R. The other two 2–2–2 tank engines by Murdoch & Aitken, their numbers 7 and 8, built in 1837 at a cost of £1,150 each. When the railway changed over to horse traction these engines were sold for twenty guineas each.

The next few years witnessed considerable activity in develop-ing railways in the northern kingdom. The year 1840 saw the opening of the Glasgow, Paisley & Greenock and the Glasgow, Paisley, Kilmarnock & Ayr Railways. The Wishaw & Coltness followed in 1841 and the Edinburgh & Glasgow in 1842. To the north, the Arbroath & Forfar was opened in 1839.

When the G.P. & G. extension reached Greenock in 1841, work-shops were established there whilst the Ayrshire line set up their shops at Cook Street, Glasgow. The former railway was engineered by Joseph Locke and the latter by Grainger & Miller who also built the Edinburgh & Glasgow.

These engineers were responsible initially for the necessary motive power of the respective undertakings and, as was the pre-valent custom, obtained the locomotives from firms who built to their specifications. Many firms were attracted by the possibilities of financial gain and increased reputation to be obtained from the new industry, and almost any engineering concern which con-sidered it had the necessary expertise indulged in the building of locomotives for one or more of the railways already constructed or envisaged. Among these firms were names which later became very well known in the locomotive building world as, for example, Andrew Barclay & Sons, Kilmarnock, who built their first loco-motive in 1840; Hawthorn & Co. Leith, an offshoot of the New-castle firm which opened the subsidiary factory in order to avoid having to ship locomotives by sea from Newcastle to Scotland. After the opening of the Royal Border Bridge in 1850 the Newcastle firm sold the Leith works to an independent company who as Hawthorn & Co. continued to build locomotives until the middle eighties. Railways which obtained their motive power from this company included the Inverness & Nairn; Inverness & Aberdeen Junction; Deeside; Perth, Almond Valley & Methven; Scottish North Eastern; and even the London, Chatham & Dover.

Sharp, Roberts & Co., founded in Manchester in 1833, became Sharp Brothers in 1843 and Sharp, Stewart & Co., in 1852 and removed to Glasgow in 1888 and took over the Clyde Locomotive Co. of 1886 later becoming part of the immense North British Locomotive Co. formed in 1903. The other partners in this undertaking were Neilson Reid & Co. who as Neilson & Mitchell supplied their first product to the Garnkirk & Glasgow Railway in 1843, and the third member of the firm was Dübs & Co. founded in 1864 and whose works were in Govan, the other two being adjacent to each other in Springburn.

That Scott, Sinclair & Co. of Greenock should be called upon to build for the Caledonian and Scottish Central Railways is not surprising seeing that from the formation of the former Robert Sinclair, a nephew of Sinclair of the Greenock firm, had been appointed general manager and locomotive superintendent.

Besides those already mentioned there were some other engineering concerns which built locomotives for railway companies and for ironworks etc. in Scotland and other countries. At the same time a large number of locomotives came from such English firms as Vulcan Foundry and Jones & Potts, Newton-le-Willows; Bury, Curtis & Kennedy, Liverpool, and George England & Co., London, among others.

John Miller, one of the partners of Grainger & Miller, was responsible for one of the early designs of 2–2–2 locomotive for the G.P.K. & A. and four were built by each of the firms, Kinmond, Hutton & Steel; T. Edington & Son; and Stark & Fulton. Some Bury 2–2–0 and 0–4–0 engines made up the initial stock in the first two years. As the eighteen-forties advanced more engines were added mainly of similar types and including some 0–4–2 and 2–4–0, the latter by Hawthorn & Co.

The Edinburgh & Glasgow started with an order to Bury for ten of the usual pattern of 2–2–0 and 0–4–0, and an order to Hawthorn for seven singles and three 2–4–0s. Both the Hawthorn types had 13×18 in. cylinders; the singles had 5 ft. 6 in. driving wheels, and the 2-4-os had 5 ft. coupled wheels. Seven engines from each firm were delivered for the opening of the line, the remainder following later. Some of these locomotives lasted so long that they were not scrapped until the final years of Wheatley's term of office which expired in 1874, long after the E. & G. had been amalgamated with the North British.

The appointment of William Paton in 1842 as Superintendent of Locomotives established a sound basis for the running of the E. & G. traffic. Unfortunately the locomotives were sadly over-worked and, in consequence, under-maintained, and were not too good to begin with so Paton had a difficult time trying to run his department efficiently. In 1844 he produced his heavy tank loco-motive for hauling trains up the Cowlairs incline, where hitherto rope haulage had been the order. *Hercules* was a six coupled loco-motive with wheels 4 ft. 3½ in. diameter. The cylinders were 15½×25 in. The middle pair of coupled wheels was fitted with a handbrake operated from the footplate, and a form of steam brake was applied to the trailing wheels. To mitigate the effect of the greasy rails in the Cowlairs tunnel hot water jets were provided in front of the wheels and cold water jets behind them. On either side of the smokebox there were sandboxes operable by the fire-man. A small amount of coke was carried in a box on the footplate and 200 gallons of water, enough for two return trips between Queen Street and Cowlairs, was contained in a small tank under the smokebox. The steeply inclined cylinders had the valve chest on top and drove on to the middle axle. Two safety valves were fitted, one on the dome and the other on the top of the firebox; these were of the spring balance type. The success of this, the first locomotive to be constructed in Cowlairs works, is shown by the fact that it hauled a 54-ton train of 12 coaches up the incline at a speed of 15 m.p.h. Later the same year a second engine of the same type was built. It was named *Samson* and whilst generally similar to *Hercules* had 16½ in. diameter cylinders and the safety valves were different. The latter were spring balance type as before, but one was mounted direct on the front ring of the boiler while the other was as before on top of the firebox. The wheels of *Samson* were 4 ft. 9 in. diameter.

The Caledonian Railway, which commenced operations on

February 15, 1848, was engineered by Locke who adopted Allan designs emanating from Crewe. Moreover, on the G.P. & G. there was Robert Sinclair, one time associate of Allan and Buddicom at Edge Hill. The influence of Allan in Scotland will be fully dealt with in the next chapter.

Several other companies which were absorbed into the Caledonian network, e.g. Scottish Central, Scottish North Eastern, and Scottish Midland Junction Railways, possessed a variety of locomotives amongst which were to be found examples built by Jones & Potts, Tayleur & Co. (later Vulcan Foundry), Scott & Sinclair and the Caledonian works at Greenock. These were Allan 'Old Crewe' type singles and 0-4-2 goods engines with single frames and underhung cylinders. Three of the companies mentioned, C.R., S.C.R. and S.M.J.R., had no less than 115 Crewe type singles between them. The Aberdeen Railway possessed a number of locomotives by Hawthorn and by several other builders, and became part of the Scottish Midland Junction in 1855 and in 1856 was absorbed into the S.N.E.R. forming the most northerly section of that great trunk line stretching from London to Aberdeen which had been envisaged by the promoters of the Scottish lines and their English counterparts.

On the North British Railway, which was opened from Edinburgh to Berwick on June 18, 1846, Robert Thornton was locomotive superintendent and he had gathered together a number of Hawthorn types rather similar to those that Nicholson was using on the Edinburgh, Perth & Dundee Railway. It has been said that the locomotive stocks of these two Scottish railways and their, later, east coast associates, the Great Northern and the York, Newcastle & Berwick, showed considerable similarities. Many of these early locomotives had very long lives, some being rebuilt more than once and not finally reaching the scrap heaps until the eighties and nineties. Equally certain is the fact that money was always tight and severe economies had to be exercised. In some instances the results were disastrous, as in the case of the E. & G. smash at Gogar in 1845, when a passenger was killed. In this case the engine, a Bury four-wheeler named *Napier*, was in a bad state of repair and had given a lot of trouble due to boiler and tube leakage. In some cases trouble arose from the poor state of the permanent way. Thornton was troubled by this on the N.B. where there was a serious outbreak of fractured axles. In trying to cure the trouble Thornton unwisely removed the inside bearings from the driving axles of his double-framed engines and merely shortened the already brief life of the crank axles.

By the eighteen-fifties it had become very evident that the small early types of locomotives such as the Burys and Planets were totally incapable of handling the increasing traffic on the ever-growing system of railways in Scotland. Moreover these four-wheelers were by no means good riding machines, particularly those having outside cylinders. The change to inside cylinders, attributed to a suggestion made by Richard Trevithick to Robert Stephenson, appears to have been founded on a desire to improve the engine thermodynamically, an improvement in the riding qualities resulting rather as a bonus. Certainly the search for increased power called for longer boilers and consequently required additional wheels. The coupling of driving wheels was generally applied to locomotives kept mainly on goods workings, passenger locomotives usually being of the 2-2-2 type. This type became very popular and developed from Stephenson's 'Patentee' type built in 1833 for the Liverpool & Manchester Railway. As in the 'Planet' type, sandwich frames were employed consisting of oak or ash planks strengthened outside and inside by iron plates. Bury, however, preferred to use bar frames. In these, instead of plates and wooden planks, the frame was built up of a series of iron bars something in the nature of a truss. This kind of frame had the advantage of lightness compared with the sandwich frame but lacked the rigidity necessary to resist the disturbing forces set up when the engine was in motion.

Bury's little engines were light and cheap and therefore popular. Whilst they had no great power they could be, and were (particularly on their own line, the London & Birmingham), used in multiple—three and even four heading one train was a frequent occurrence. They suffered from lack of heating surface, particularly in the D-shaped firebox. At first Bury used iron plates for the fireboxes but changed to copper some years after Stephenson had done so. Although the Bury engines showed a definite advance in locomotive design, the 'Planets' were an even greater milestone in the history of the steam engine.

Boiler construction posed some considerable difficulties in

those days. About 1831 as many as 22 plates would be required in the making of one boiler, so small was the size of plate available. This meant a very large number of riveted joints, lap joints at that, and it was not until the ironfounders started to produce larger plates that a reduction in the number of plates, joints and rivets could be effected. By 1838 however, so much advance had been achieved that a barrel could be made from four plates, each the full length of the barrel, e.g. between seven and eight feet.

Firebox crown stays were of simple construction in the early engines, being merely angle-iron stiffeners, but by the middle thirties iron girders were taking their place. In his earliest engines which had dome-shaped crowns, Bury either used palm stays or no stays at all; later, in those engines which had the D type of firebox, he used shorter girder stays. Iron was the usual material of which side stays were made and again, by the middle thirties, the Stephensonian practice of using copper became more general.

The position of the steam dome on the boiler was the subject of much debate and two schools of thought dominated the locomotive world for some years; just as, some time later, there were to be heated arguments on the question of boilers with domes versus domeless boilers, so there was a difference of opinion as to whether the dome should be over the firebox or placed well forward on the barrel. In favour of the latter practice it was held that if so placed, the dome, which was in reality the main point from which steam was collected, was as far as possible from the point of most violent ebullition, i.e. the firebox. Secondly, the length of steam pipe required was shortened resulting in increased efficiency of glands and joints. Thirdly, the short steam pipe occupied less space in the boiler, a matter of no inconsiderable importance in these small boilers.

Against these arguments there was one main objection: placing the dome so far forward tended to put excessive weight on the leading end of the engine. Some builders, e.g. Mather Dixon & Co. and Tayleur & Co., employed two domes, one on the firebox and one well forward on the barrel just behind the chimney, and Hawthorns struck a compromise and placed the dome near the middle of the barrel, which has been the designers' usual choice ever since. Illustrating these divergences of opinion were the 2–2–2s designed by Miller for the G.P.K. & A.R. and built by

Edington; Kinmond, Hutton & Steel; and Stark & Fulton in 1840-1. All had the dome well forward on the boiler barrel, while the Jones & Potts singles built in 1848 for the s.c.r. had domes on the fireboxes.

The functions of the dome were twofold. First, it served to increase the steam space, and second, it usually housed the regulator valve and because of its height ensured that the driest steam was collected. Those engines having two domes had the regulator valve in the front one.

It was early appreciated that some means would have to be provided to prevent loss of heat by radiation and to this end boiler barrels were usually lagged with wooden battens, generally mahogany, polished on the outside. The firebox was often left uncovered until later, when firms began lagging that also. In the early forties it was common to insert a layer of felt between the wood battens and the boiler plates but the practice was discontinued when it was found that rain seeping between the joints caused corrosion of the plates.

An excellent example of early locomotive design is afforded by some 2-2-2 engines built by the Vulcan Foundry, Newton-le-Willows, in 1847 for the Caledonian Railway. These engines followed the general Allan pattern, the two outside cylinders, 15×20 in., being secured in the usual 'Old Crewe' style with the steam chests fitted in slots cut in the inside frames. The smokebox wrapper plate was extended downwards to embrace the cylinders which were inclined at an angle of 1 in 9 to the horizontal. The piston rods, crossheads, slidebars and connecting rods were between the outside and inside frames and the latter extended the full length of the engine and did not stop short in front of the firebox as in some early types of double framed engines. The Stephenson link motion was between the inside frames and drove the valves through a rocking shaft above the leading axle. The slide valves were arranged vertically and separate steam and exhaust pipes led to and from each cylinder. The exhaust pipes joined together below the base of the chimney to form the blast pipe the orifice of which, varying from $3\frac{11}{16}$ to $4\frac{1}{32}$ in. diameter, was level with the bottom of the chimney. A 'T' piece on the smokebox tubeplate allowed the internal main steam pipe to bifurcate, each $4\frac{3}{4}$ in. diameter branch leading direct to the rear end of one of the steam chests. The main steam pipe extended the full length of the boiler and firebox and terminated at the door- or backplate of the latter in a stuffing box which contained a spindle to operate the regulator valve. A steam dome surmounted the firebox and a large diameter pipe inside the dome led down to a junction with the main steam pipe, which at this point contained the regulator valve. This was a plug type valve which fitted the internal diameter of the steam pipe and was attached to the spindle which passed through the stuffing box on the backplate of the firebox. On the outer end of the spindle was fastened a handle which could be moved across a quadrant and by means of the angle to the firebox at which the latter was set, pull the valve back to admit steam from the boiler to the steam chests, or by reversing the movement of the handle to shut off the supply of steam from the boiler.

The boiler barrel was 3 ft. $6\frac{3}{4}$ in. diameter and contained $158 \times 1\frac{3}{4}$ in. tubes, which at 10 ft. $1\frac{1}{2}$ in. length between tubeplates, gave 732·9 sq. ft. heating surface. The firebox was 3 ft. long, 3 ft. 6 in. wide and 3 ft. $5\frac{1}{2}$ in. from firebars to crownplate giving 49 sq. ft. The grate area, 10·5 sq. ft., comprised 24 firebars and the air space through them was equal to approximately 45 per cent of the grate area. The ashpan was closed at the back end and apparently did not have a damper door at the front. Two safety valves of the Salter type were mounted, one on a manhole cover midway along the barrel, the other on top of the dome. The working pressure was 90 lb. per sq. in.

Two feed pumps were bracketed to a cross stay immediately in front of the firebox. The pump rams were operated by connecting rods attached to lugs on the backs of the back gear eccentric straps and water was delivered to the boiler through clack valves mounted on the firebox sides about 6 in. behind the tubeplate. The feed was obtained from the tender water tank through ball and socket jointed pipes under the footplate. The tender was four-wheeled and carried 800 gallons of water in a horseshoe-shaped tank. There was space for two tons of coke or three tons of coal between the horns of the horseshoe. The intermediate coupling between engine and tender was an arrangement of two 'D' links attached to a central link by four trunnion bearings. The front 'D' link was anchored by a pin passing through the engine dragbox. The rear

45

'D' link was secured by a similar pin which also passed through the eye of an extension of the buckle of a laminated spring in the tender dragbox. Side safety links were also fitted. At the rear of the tender another laminated spring with an extended buckle provided the drawhook. In addition there was a four-link chain.

The leading and trailing wheels were 3 ft. 6 in. diameter and had outside bearings with underhung laminated springs to the former and overhung springs to the latter. The driving wheels were 6 ft. diameter and underhung laminated springs were provided. The bearings were in the inside frames. The tender wheels were also 3 ft. 6 in. diameter and had overhung laminated springs. It is noticeable that all springs were reasonably accessible.

There was no brake on the engine but all four tender wheels were fitted with wooden blocks operated by a handwheel mounted on the right-hand horn of the tank. The wheel was attached to a long shaft screwed at the end and working in a fork the lower end of which moved a cross-shaft between the axles. Simple rigging applied the blocks to the back of the leading wheels and the front of the trailing wheels.

The buffers were round leather pads of large diameter reinforced by iron bands.

No protection from the weather was afforded to the enginemen either in the form of a cab or simple weatherboard.

The weight of the engine is given as 19 tons and of engine and tender together in working order 28 tons.

Clark records these engines as running the $27\frac{1}{2}$ miles from Edinburgh to Carstairs in 32 minutes, an average of 51 m.p.h., but adds that time was rarely kept with trains of four coaches weighing 35 tons. However these engines ran for many years, the last in service was scrapped in 1874.

One of the difficulties encountered in maintaining locomotives at this time was due to the arrangement of inside cylinders with a common valve chest between them. Thus valve setting was difficult and maintenance not made easy by having the valve motion close to the engine centre line between the connecting rods. In an endeavour to overcome these awkward circumstances some 2-2-2s built in 1848 were designed with the cylinders inside, but the steam chests outside, projecting through the frames which were cut out for the purpose. The Stephenson's link motion was arranged outside the frames and was thus more readily accessible. The main criticism of this design was, of course, that there was loss of heat by radiation from the outside steam chests. Another interesting feature of this design was that the bosses of the driving wheels were considerably enlarged so as to form the outer webs of the cranks. These engines were built by Vulcan Foundry and the boilers were about 12 ft. 8 in. between tubeplates, with domes on the front rings and raised fireboxes. The latter were of an unusual shape; the front half of the 3-foot-long grate was the full width possible between the outside frames and the rear half was narrower so as to accommodate the trailing wheels.

In 1846 the Whitehaven firm of Tulk & Ley had built two engines to Crampton's design for the Namur & Liege Railway and, in the following year, another was built for the Dundee, Perth & Aberdeen Junction Railway. It followed the classic Crampton style in having the 7 ft. driving wheels in rear of the firebox and two pairs of 3 ft. 9 in. carrying wheels in front. The cylinders were 16×20 in. and the eccentrics of the Gooch valve gear were mounted on return cranks outside the connecting rod big ends. The boiler centre line was very low, only 4 ft. 7 in. above rail level, in an endeavour to keep the centre of gravity as low as possible. The tube heating surface was 927 sq. ft. and that of the firebox 62 sq. ft.; the grate area was $14\frac{1}{2}$ sq. ft. The working pressure was 50 lb. per sq. in. and the total weight 24 tons. The springing of this engine was unusual: five point suspension was provided by springing each of the carrying wheels separately, and the driving wheels by a transverse laminated spring, the buckle of which formed part of an iron casting attached to the firebox. Crampton was probably the first British locomotive engineer to appreciate the behaviour of a locomotive on a curved track and it was from his study of these characteristics that he developed his design believing that the manner of springing, low centre of gravity, and the position of the driving wheels would produce the improvement in the riding of a locomotive that he sought. One serious disadvantage was that the adhesion weight was very far in rear of the centre of gravity. Nevertheless, these engines proved speedy and comparatively steady runners. Rosling Bennett records that this locomotive was sent from Whitehaven to Hull under its own steam via the Sheffield, Ashton and Manchester route over the Pennines. From Hull it was shipped to Leith. There was at this date, August 1847, no through rail connection from Carlisle to central Scotland. This was Tulk & Ley's 14th engine and was given the name *Kinnaird* after the chairman of the D.P. & A.J.R.

E. B. Wilson of Leeds were also constructing locomotives to Crampton's designs and supplied one to the North British Railway in 1848 and two to the Aberdeen Railway. While following the general arrangement of *Kinnaird* very closely there was one notable difference, the leading wheels were 4 ft. 6 in. diameter, the second pair 3 ft. 6 in. and the driving wheels 6 ft. whilst the cylinders were 16×20 in. When 19 years old this locomotive underwent considerable rebuilding at St Margaret's works in Edinburgh, when the original intermediate wheels were replaced in the trailing position and the driving wheels put in their place. At the same time the alterations to the boiler all helped to give this engine the appearance of a 'Jenny Lind'. Later in life it was again rebuilt, by Drummond who gave it a domed boiler and a cab. The running number on the N.B.R. was 55, though it was eventually renumbered 1009. When Queen Victoria opened the Royal Border Bridge at Berwick in 1850, No. 55 hauled the special royal train. For this occasion it was turned out from St Margaret's works painted in the Stewart dress tartan as a compliment to Her Majesty. The two Cramptons for the Aberdeen Railway were built in 1850, the boilers had domes and the intermediate wheels had inside bearings only.

One of the small tank engines of the 'Little England' type built by George England & Co. of Hatcham Iron Works, London, was sent to the Edinburgh & Glasgow Railway in 1850. The makers guaranteed that it would successfully work express trains of seven coaches between the two cities and maintain good time on a fuel consumption, not exceeding 10 lb. of coke per mile. If the experiment was successful, the railway company would purchase the locomotive for £1,200, but if unsuccessful England & Co. would remove the locomotive and pay the costs of the trial.

In the event the coke consumption worked out at 8 lb. 3 oz. per mile as compared with 29 lb. 1 oz. for *Sirius*, one of two engines designed by Paton and built at Cowlairs in 1848. The England engine, named *Little Scotland* was running up to 95 miles per day and frequently exceeding 60 m.p.h. It was a 2-2-2 weighing only 13 tons. The cylinders were 9×12 in. and the driving wheels

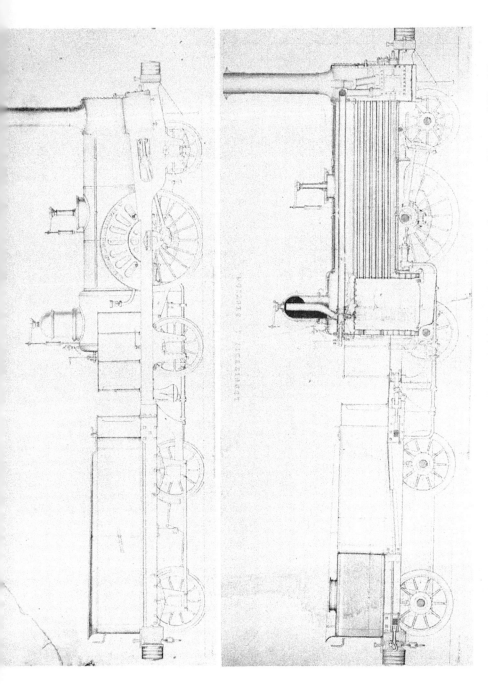

5. *Above:* Caledonian Railway, 2–2–2 No. 15, Vulcan Foundry, 1848—elevation.   *Liverpool City Museum.*
6. *Below:* Caledonian Railway, 2–2–2 No. 15, Vulcan Foundry, 1848—longitudinal section
   Liverpool City Museum.

CALEDONIAN RAILWAY

PLAN OF ENGINE WITH CYLINDRICAL PART OF BOILER REMOVED

PLAN OF TENDER

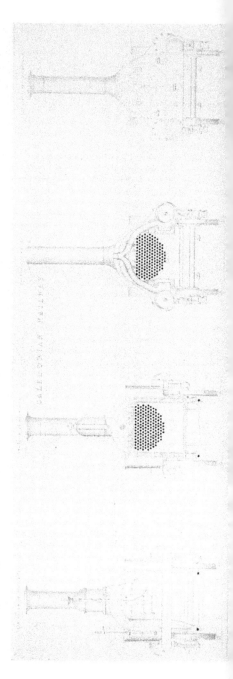

7. *Above:* Caledonian Railway, 2–2–2 No. 15, Vulcan Foundry, 1848—plan. *Liverpool City Museum.*

8. *Below:* Caledonian Railway, 2–2–2 No. 15, Vulcan Foundry, 1848—cross elevations and sections. *Liverpool City Museum.*

CALEDONIAN RAILWAY

9. *Above:* Great North of Scotland Railway, 2–4–0 No. 4, Clark, 1854.
Builders, Fairbairn & Co.
Railway Magazine.

0. *Below:* Great North of Scotland Railway, 4–4–0 No. 29, Cowan, 1862.
Built by R. Stephenson & Co.
Railway Magazine.

11. Caledonian Railway, 2–4–0 158 Class, Beyer Peacock, 1861. Order No. 415.
By permission of Manchester University.

4 ft. 6 in. diameter. The position of the cylinders was unique as they were placed far behind the leading axle about one third of the distance along the frames. The latter were of the sandwich type, single, outside the wheels. Coke and water sufficient for 50 miles were carried and there was a primitive feed water heater in the smokebox. The reputation of these engines for steady running was attributed to the low centre of gravity and the long wheelbase. Other engines of the type found their way to the Caledonian and the Dundee, Perth & Aberdeen Junction Railways in 1852–3.

The foregoing brief account of the locomotive history of the first two decades of Scottish railways brings the narrative to the stage of development when the main workshops of the railways were established under the supervision of the locomotive superintendents whose names were to become justly famous throughout the world.

By no means all the early locomotives have been described, nor have all the small companies or representatives of their locomotive studs been mentioned. For example the Pollok & Govan and Portpatrick Railways had engines of their own as had the Wishaw & Coltness and several other companies of the same period. In general their engines were very similar to those types which have been discussed above.

Chapter III

THE TREND-SETTERS

By the middle of the eighteen-fifties the railway industry in Scotland was becoming more firmly established, particularly as regards the locomotive departments. On the Glasgow & South Western, Patrick Stirling had been installed as locomotive superintendent on May 2, 1853, in succession to Peter Robinson, who had managed the department through the early years of the Glasgow, Paisley, Kilmarnock & Ayr and the amalgamation with the Glasgow, Dumfries & Carlisle. Robert Sinclair of the Glasgow, Paisley & Greenock had been appointed to a similar position on the Caledonian, and was general manager of the company also. In addition, Sinclair was supervising the locomotive department of the Scottish Central Railway and continued to do so until September 1853, when these duties were taken over by Alexander Allan who came to Perth from Crewe. Robert Thornton had left the North British and had been succeeded, first by Smith who resigned after two years, and then by the Hon. Edmund Petre who came from the Shrewsbury & Birmingham. Petre's tenure of office was even shorter and he was dismissed in 1854 for intoxication. He was followed by a Scotsman, William Hurst, who had been born at Markinch, Fife, in 1810 and was previously outdoor superintendent of the Lancashire & Yorkshire. The Great North of Scotland had appointed Daniel Kinnear Clark who divided his time between Kittybrewster and an office in London from which he wielded some form of remote control.

In 1855 the small companies which ultimately went to the making of the last of Scotland's major railways, the Highland, appeared. These were the Inverness & Nairn, Inverness & Aberdeen Junction, Inverness & Perth Junction and the Ross-shire Railways, of which William Barclay was in charge of the locomotive departments. He was a nephew of Alexander Allan. Lastly, at Cowlairs, William Paton was in full stride endeavouring to pull his department out of the doldrums and to rehabilitate himself after his year in jail following the unfortunate Gogar accident.

Patrick Stirling, at the Cook Street works in Glasgow, was finding them too cramped for his expanding responsibilities. Accordingly, the G. & S.W.R. bought land at Kilmarnock and built a new establishment there. It was opened in 1856 and the first locomotives constructed in these new workshops were Stirling's first real attempt at express locomotive design, his No. 2 class of 2–2–2, of which twelve were built between 1857 and 1860.

The Caledonian, too, was finding its works not only cramped but too remote, tucked away at the end of the Greenock line. A new site was found in Glasgow itself, at Springburn, and there the company built the St Rollox workshops which also opened in 1856. Sinclair only remained with the Caledonian Railway long enough to get the new works established when he resigned, to take up a similar post on the Eastern Counties Railway in England. His successor at St Rollox was Benjamin Conner who came down the road from Neilson & Co. where he had been works manager.

Of the half-dozen or so locomotive superintendents in Scotland at this time two in particular were to leave a very decided mark on the locomotive history of that country, Patrick Stirling and Alexander Allan. Both were Scotsmen and both had been south of the border to gain experience during the earlier parts of their careers.

A son of the manse, Stirling was born at Kilmarnock on June 29, 1820, and when he was fifteen years of age he was sent to Dundee, to be apprenticed to his uncle, whose firm Stirling & Co. was engaged in building locomotives as one of its activities. In 1843 the young man went to Robert Napier to gain some marine experience. Three years later he joined Neilson & Mitchell, afterwards Neilson & Co., at their Hyde Park works in Springburn and rose from journeyman to foreman. Returning to railways he had a short spell as locomotive superintendent of the Caledonian & Dunbartonshire Railway, a very small concern physically un-connected with any other railway but affording a link between the Clyde steamers and those on Loch Lomond. This did not last long for he next went to Laurence Hill at Port Glasgow, again on marine work and then to R. & W. Hawthorn at Newcastle. This last association was to have a deep and lasting effect on the young engineer as will be seen when his locomotive designs are considered.

Thirteen years Patrick Stirling served the G. & S.W.R. and they paved the way to his triumphal successes on the G.N.R. in later years but his first efforts appear, at this late age, to have been somewhat fumbling. His first G. & S.W. engines were four outside cylinder 2–2–2s built by Neilson & Co. in 1855. In the same year he ordered from Hawthorn & Co. four odd looking 0–4–0 intermediate crankshaft locomotives with a 12 ft. 1 in. wheelbase. These were followed by two 0–6–0 outside cylinder engines which, again, Hawthorn built. They were not popular and, like the 0–4–0s, had only a short life. At this time Beattie on the L. & S.W.R. was experimenting with coal in search of a more economical fuel than the ubiquitous and expensive coke. Thus, in 1857, the G. & S.W.R. placed an order with Beyer Peacock & Co., Manchester, for a 2–4–0 engine to Beattie's specification. This locomotive had two cylinders 16 × 22 in., coupled wheels 5 ft. diameter and the heating surface was 694 sq. ft. from 371 × 1¼ in. tubes, the firebox having 120 sq. ft. and the grate area being 16 sq. ft. The boiler pressure remains unknown today but may well have been 130 lb. per sq. in. which was the pressure used by the Midland on similar types about this time, and for whom Beyer Peacock also built a Beattie type engine, and by Beattie himself in 1859. This engine had the Allan straight link valve motion.

The total grate area, 16 sq. ft., was in fact divided into two portions. The firebox was thus a double one, a water bridge, something on the lines of the modern thermic syphon but not performing the same function, separating the two portions. Coal, fired through the lower of two fireholes, was burned in the back grate and the gases passed over the incandescent coke fire in the front part of the box completing the process of combustion. Coke was fed through the upper firehole. The name *Galloway* was bestowed on the engine which was No. 109 in the G. & S.W. stock. The Beattie system was very efficient and brought the coal consumption down to around 24 lb. per mile and the amount of coke consumed was small. An evaporation rate of over 8 lb. water per lb. coal was achieved.

Stirling's first real express locomotive, and the first to be built in the new Kilmarnock establishment, was a 2–2–2, with 16 × 21 in. cylinders and 6 ft. 6 in. driving wheels. No other dimensions have survived but it seems evident from such illustrations as exist that the design was compounded of elements of the family products of Dundee and of Peter Robinson. The boiler carried a dome, as did ten 0–4–2s which were built by Hawthorns in 1858. In 1860 there appeared the '22' class of 0–4–2 of which 20 were built by Sharp Stewart by 1862. These were notable for two innovations: they were not fitted with domes and one of the class was the first in Scotland to have a Gifford injector.

It would seem that, although Stirling had been acquainted with the design of boiler which did not incorporate the use of a dome, his experience in this field being during his service with Hawthorn & Co., he had not previously made up his mind regarding the use or non-use of this feature. However, it is significant that, having produced a successful engine with a domeless boiler, he stuck to this policy throughout the remainder of his career. Since coming to the G. & S.W.R. he had had an opportunity of studying the performance of both types of boiler, there being the domeless Hawthorn of 1851 to compare with his own designs in their everyday work

In all Stirling produced a further seven classes of locomotive for the G. & S.W. in the years which followed until his departure from Kilmarnock. In 1860 a domeless boiler edition of the '2' class was turned out from the company's own works. Thus, the '40' class 2–2–2 had a long, useful life of some twenty years first on express work, for which they were designed, and later on branch lines and as pilots of expresses. Cylinders and driving wheels were similar to their predecessors but they were $1\frac{1}{2}$ tons heavier at $24\frac{1}{2}$ tons. The earlier engines had four-wheeled tenders of 1,200 gallons capacity while the later ones had six-wheeled tenders of 1,500 gallons capacity. These tenders were all to Stirling's own design and were wooden framed. The tank was in the shape of a horseshoe with the bearing springs behind the side plating, the springs being somewhat difficult of access. Later engines of the class were fitted with the design of cab which Stirling produced in an effort to afford some protection to the enginemen. These early cabs were somewhat narrow and were provided with a circular porthole on each side as a lookout. Later, because of complaints from drivers that they could not see out of these ports sufficiently well, the side sheets and top of the cab were cut back above the waist and became the standard G. & S.W.R. cab arrangement for many years and was also copied by other railways.

One other class of 2–2–2 was built during the Stirling régime and it proved to be a considerable advance on the previous designs. The cylinder diameter remained 16 in. but the stroke was increased to 24 in. and the driving wheels to 7 ft. At 785 sq. ft. and 85 sq. ft. the heating surfaces of tubes and firebox were slightly smaller than before, but the boiler pressure was lifted to 125 lb per sq. in. and the weight of the engines to 28 tons 8 cwt. Eleven of these, the '45' class, appeared between 1865 and 1868 and proved to be fast running and efficient machines.

Of the 0–4–2 type two further classes were built, one, the '131' class, by Hawthorn & Co. in 1864, was the last to be built by this firm for the G. & S.W.R.; the other was the '141' class by Neilson & Co. in the following year. The Hawthorns were very similar to the '32' class of 1858 but, of course, had domeless boilers. Neilson & Co's engines were 1 in. larger in the diameter of the cylinders and the coupled wheels were the first engines on this railway to have Allan's straight link valve motion except No. 109, the 'Beattie'. Early in Smellie's term of office four of these were rebuilt as 0–4–2T for shunting and had no rear bunkers, merely a flat plate. Similarly rebuilt in 1877–8 by Patrick's brother James were eight of the '22' class. These were used on the suburban services of the City of Glasgow Union Railway, some lasting until 1908. Patrick Stirling's first inside cylinder design was brought out in 1862 and was surprisingly an 0–6–0, a wheel arrangement he did not care for; however he seems to have gained some satisfaction from its performance, for his last design, the '58' class 0–6–0, appeared in 1866. Six were built at Kilmarnock in 1866–7 and twenty by Neilson & Co. between 1866 and 1869. Dimensionally similar to the 0–4–2s of 1865, they had Stephenson's link motion instead of the Allan gear.

One other class should be mentioned. This is the '52' class 0–4–0 tender engine. Kilmarnock turned out six in 1865–6 and they were the first of the type since the dummy crankshaft engines of 1855. The later design was domeless and had a four-wheeled tender and proved useful engines in the sharply curved colliery areas.

The year 1843 witnessed the opening of the new locomotive building and repair establishment at Crewe and the removal thereto of Alexander Allan from Edge Hill where he had been foreman of locomotives under William Buddicom. Allan was a Scot who had served his apprenticeship with a Mr Gibb, a wheelwright at Montrose, Allan's native town. In 1840 he relinquished the position of works manager to Forrester & Co., of Vauxhall Foundry, Liverpool, to cross the town to Edge Hill, moving again in 1843 to Crewe. There he made a name for himself by introducing his own design of outside cylinder engine on an inside cylinder engine line, based on experience gained whilst he served Forrester & Co. The troubles of the Grand Junction Railway because of broken driving axles, and the way in which the difficulties were overcome by the adoption of Allan's front end modification, have often been told together with the way in which this model or modifications of it found their way to other companies.

Allan served the G.J.R. for thirteen years and in September 1853 was appointed locomotive superintendent of the Scottish Central Railway with his headquarters at Perth succeeding Sinclair. Sinclair and Allan were old associates, having been contemporaries at Edge Hill. Either because of his access to his old friend at Crewe, or with an eye towards some degree of standardization, since the amalgamation of the S.C.R. with the Caledonian was generally supposed to be only a matter of time, Sinclair had ordered locomotives of the 'Allan' or 'Old Crewe' type, departing from this in one respect only. That was the design and construction of an 0–6–0 instead of a 2–4–0 for goods traffic. The latter were introduced by Allan between 1857 and 1864 and were built by Fairbairn, Neilson, and Sharp Stewart. The latter firm also built a series of large 2–2–2 the first of which came out in 1864. Some similar locomotives were built at Perth but they had a slightly smaller heating surface and a wheelbase shorter by 6 in. Most of these engines were fitted with Allan's design of mid-feather firebox.

1856 was notable for Allan's presentation of his celebrated paper to the Institution of Mechanical Engineers introducing his patent valve gear. In the preceding year four 0–4–2 goods engines built by Vulcan Foundry for the Scottish Central were fitted with his gear which became extensively used. The expansion link being straight, instead of curved to the radius of the eccentric rods, simplified machining and lowered production costs, a not inconsiderable point in its favour. Allan also claimed that a better steam distribution was obtained than with the more orthodox gears.

Allan was an early engineer to use steel for fireboxes, and a good degree of success attended his use of this material in ten s.c.r. locomotives built 1860–3. The plates were obtained from Sheffield and varied in thickness from $\frac{3}{8}$ in. for tubeplates to $\frac{5}{16}$ in. or $\frac{3}{8}$ in. for side plates. Some ten years later Conner, who followed Sinclair on the Caledonian, stated that only two of the ten fireboxes had undergone slight repairs. He himself had made a similar essay in the use of steel but employed thicker plates, $\frac{3}{4}$ in. for tubeplates, $\frac{3}{8}$ in. and $\frac{1}{2}$ in. sideplates and $\frac{1}{2}$ in. backplates; all these had failed due to the development of cracks. Thus early it was established that thin steel plates gave more satisfactory results in service but that when steel plates of similar thickness to copper plates were used failure was nearly always certain.

At this period there was considerable diffidence about using steel in the locomotive industry. Krupps of Essen, through the energy and initiative of their British representative, Alfred Longsdon, were trying to get British locomotive engineers interested in the use of steel, particularly for tyres and axles. Alfred Longsdon's brother, Robert, was related by marriage to Henry Bessemer. This connection undoubtedly facilitated co-operation between the two big steel firms. Publicity material was sent out in 1858 to the British railway companies, advocating the use of tyres and axles made of Krupp's steel. Reception was somewhat mixed. Cowan of the g.n.s.r. was lukewarm and expressed the opinion that, from an economy angle, the use of steel would be more effective when used for crank axles. Since all the g.n.s.r. locomotives at that time were outside cylinder locomotives he was committing himself to nothing. On the Caledonian, Benjamin Conner was even less enthusiastic for he supposed that there might be an increased liability to slipping of the wheels which with the, to him, unusual cost of the tyres might show little advantage by a trial of them.

In December 1854, on the recommendation of Joseph Mitchell, engineer of the Inverness & Nairn Railway, Allan was appointed as consultant with full powers to plan and superintend the construction of locomotives and rolling stock. The following year he installed his nephew William Barclay as locomotive superintendent of the line and of the Inverness & Aberdeen Junction Railway. Later Barclay was also in charge of the locomotive department of the Inverness & Perth Junction, all of which combined in 1865 to form the Highland Railway. The locomotives which Allan ordered and of which Barclay had the superintendence were of his own design of 2–2–2 type for passenger services, and 2–4–0 for goods. The builders were Hawthorns of Leith. 1855–7 saw delivery of the first four which were 2–2–2 with 15×20 in. cylinders and 6 ft. driving wheels. Numbers 1 and 2 bore the names *Raigmore* and *Aldourie*. The boilers were pressed at 100 lb. per sq. in. and carried no dome. There were two safety valves of the spring balance type, one over the firebox which had a raised casing, and one midway along the barrel. Allan's straight link motion was used. The four-wheel tenders carried 1,100 gallons of water. The 2–4–0s appeared between 1858 and 1862, also from Hawthorns. They were slightly larger and incorporated a number of features in common with the 2–2–2s.

The line along the Buchan coast was comparatively flat compared with that over the mountains, with two summits, one of 1,052 ft. near Dava and one of 1,484 ft. at Druimauchdair, resulting in gradients of the order of 1 in 60 for miles at a stretch. The little singles would be of small value on such a difficult road and much of the haulage on the mountain line was given to the 2–4–0s. For this purpose Sharp, Stewart & Co., Manchester, were given an order for ten 2–4–0s of a slightly larger type than had previously been used. These were delivered in 1863 and were followed in 1864 by ten more. The first batch had 17×22 in. cylinders and 5 ft. 1½ in. driving wheels, and the later batch had the slightly longer stroke of 24 in. In each type the boiler pressure was 150 lb. per sq. in. but the heating surface of the 1864 engines was increased from 1,164 sq. ft. to 1,328·5 sq. ft.

When the amalgamation of the smaller railways to form the Highland took effect on February 1, 1865, there were 55 locomotives in service. All of these were either 2–2–2 or 2–4–0 type except two which were 0–4–0 tank engines. One of these 0–4–0T was taken over by the I. & A.J. from the Findhorn Railway which it absorbed in 1862, and was a standard Neilson tank locomotive, whereas the other was a Hawthorn inside cylinder engine built in 1863. Until Peter Drummond arrived in 1896 it was one of the few inside cylinder locomotives the company possessed.

The arduous nature of the main line south soon demonstrated the need for engines stronger than the little singles. The latter requirement was in some sense achieved by rebuilding some of the singles as 2-4-0s during the 1870s, but the former was only obtained after a managerial upheaval. Barclay was frequently absent from duty to go fishing. It was unfortunate that one of the directors should happen to see Barclay engaged in the pursuit of his hobby on a day when a serious engine failure occurred on the Perth road, and this at a time when locomotive failures were all too prevalent on the Highland. In consequence, Barclay was dismissed from the service and paid three months' salary in lieu of notice. Until a successor was appointed the department was carried on by David Jones, Barclay's assistant, an able young man of 31 who had been with him since the commencement of the I. & N.R. in 1855. On June 19, 1865, the directors appointed William Stroudley to be locomotive superintendent.

On the neighbouring, but un-neighbourly, Great North of Scotland Railway the superintendency of the locomotive department was in the hands of Daniel Kinnear Clark. An engineer of no mean calibre, Clark exercised his supervision from a London office, on-the-spot supervision was left to Clark's assistant John Folds Ruthven who was styled 'foreman of workshops'.

The first locomotives of this line were some 2-4-0s built by William Fairbairn & Co. Seven were ordered for the opening of the line on September 12, 1854. Clark supervised their construction but, what with late delivery and poor workmanship, the company passed through a difficult period until all seven were at work. In the succeeding years until 1859 four more of the same type were built. For goods work Fairbairn & Co. built five 2-4-0s in 1855, also to Clark's design, again without the standard of workmanship desired.

All these Fairbairn engines had 15 in. cylinders except the later batch of four which had the diameter increased to 16 in. The stroke was similarly increased from 20 in. to 22 in. Heating surfaces were very small, the first seven having only 686 sq. ft. of tube surface

58

and 63 sq. ft. in the firebox, the later engines 898 sq. ft. and 68 sq. ft. respectively, while the goods engines had 750 sq. ft. and 58 sq. ft. A grate area of 10·5 sq. ft. made these locomotives considerably inferior dimensionally to many of their contemporaries though they carried the comparatively high boiler pressure of 130 lb. per sq. in. The lengthening of the stroke of the goods engines was effected by the use of eccentric crank-pins but these entailed so much trouble due to fractures in service that later there was a reversion to the original 20-in. stroke.

These troubles together with the delay in delivering the locomotives from the builders and Clark's remote control culminated in a first class row. A condition of his employment was that Clark should reside in Aberdeen but he claimed that this was inimical to his advancement in his profession and held that a good assistant on the spot could carry out all that was necessary. The board disagreed and the upshot was that Clark resigned his post as from July 1855, from which date his assistant, J. F. Ruthven, was appointed at a salary of £160 per annum.

Money was always tight and repairs, however costly, were very necessary and Ruthven had a difficult time fighting for every penny he needed. He only contributed one design of locomotive to the G.N.S. stock, an 0–4–0 well tank, of which two were built by Beyer Peacock & Co. in 1856. These little engines were originally intended for banking trains from the Waterloo terminus to Kittybrewster and had 15 × 24 in. cylinders and 4 ft. 6 in. wheels. The boilers were comparatively large for the type of engine and the period. The barrels were 11 ft. long and 3 ft. 7¾ in. diameter. The heating surface was 890 sq. ft. The grate area was 11·1 sq. ft. and the working pressure 130 lb. per sq. in. A tank beneath the boiler had a capacity for 336 gallons of water and a small bunker at the side of the firebox carried 4½ cwt. of coke. They weighed 24 tons of which 14 tons rested on the driving axle. These rather ugly machines did a vast amount of work mainly at Aberdeen. When reboilered in 1887 they were given cabs in place of the weatherboards which had been provided a few years after their first appearance. In 1915–16 they were sold to the government for War service, after which one passed to a contractor and survived for a few more years.

The G.N.S. board appears to have indulged in quarrels not only with their contemporaries and neighbouring concerns but also among themselves about their own officers so that, once again, in the space of a few years, a new locomotive superintendent was required, this time because Ruthven resigned over the question of repair costs; the year was 1857.

The new chief was William Cowan who had been works manager since 1855. Cowan had been born in Edinburgh in 1823 and when 16 years old went to serve his apprenticeship in the Arbroath works of the Arbroath & Forfar Railway. He then spent a number of years with the Edinburgh & Glasgow and Great Northern Railways, returning to Scotland in 1854 at the age of 31. In September 1854 he joined the Great North of Scotland Railway being promoted works manager the following year.

On succeeding Ruthven, Cowan's first design of locomotive was a 2–4–0 slightly larger than Clark's engine. It appeared in 1859 and the builders were Robert Stephenson & Co. Three years later the same firm delivered the first passenger locomotives in Scotland to have a leading bogie and four coupled wheels. Again comparatively small, these engines had cylinders only 16 × 22 in. and coupled wheels 5 ft. 1 in. diameter. The tube heating surface was 902 sq. ft. while the firebox contributed 70 sq. ft. The grate area was only 10·2 sq. ft. and the boiler pressure 120 lb. per sq. in.

In this class the main interest centres on the bogie, which in Scotland at any rate, was a new feature. Apparently there had been some conversation between Cowan and J. D. Wardale, Robert Stephenson & Co.'s chief draughtsman, on the subject and, at the time when the design of the G.N.S.R. locomotives was being prepared, Wardale was engaged on a new design of bogie tank engine for the Metropolitan Railway; it had springs instead of inclined planes for side control, and sufficient side-play to enable it to pass round the sharpest curves without grinding, and that side control was provided for by means of a pair of volute springs.

Wardale's bogie was, in fact, never constructed. It was the typical short wheelbase type of bogie of the day. The bogie of the Stephenson engines of 1862 was a different matter. This had a long wheelbase, consistent with outside cylinders, and the frames at the front end of the locomotive were cut out to allow pivotal freedom to the bogie. Nine engines, Nos. 28–36, were built between 1862 and 1864. At least one of them had unusual springing of the bogie; over each bogie axlebox there was a laminated spring and on each side of the bogie the inner spring pillars were attached to an equalising beam pivoted at its centre.

So satisfactory did these engines prove, more especially on the unfinished Speyside lines which abounded with sharp curves, that Cowan's design concepts were strengthened and he adopted the 4–4–0 wheel arrangement as his standard for the G.N.S.R. main line locomotives. It remained so for the rest of that company's separate existence.

Improvements in bogie design very naturally followed and to William Adams of the North London Railway must be given the credit for developing this form of truck to its modern form. His first long wheelbase bogies had lateral play but lacked any side control. Later, indiarubber check springs were introduced to provide this feature and, later still, a circular brass pad was provided at the bogie centre to take the front end weight and introduce a degree of friction in the manner recommended by Fernihough of the Eastern Counties Railway some twenty years earlier. Other firms were quick to take up the new development and Neilson & Co. who built the next of Cowan's engines incorporated an improved bogie with inside frames and springs.

These 1866 locomotives by Neilsons differed from their predecessors in several respects. The cylinder diameter remained 16 in. but the stroke was increased to 24 in. and the coupled wheels to 5 ft. 7 in. The heating surface and grate were also increased slightly, though the latter was still only 13 sq. ft. The working pressure was raised to 140 lb. per sq. in. Six engines of this class were turned out, Nos. 43–48, and some were still in service when the 1923 amalgamations took place.

Cowan's designs were neat and workmanlike. The 2–4–0s and 4–4–0s had raised top fireboxes with large domes thereon, the boiler barrel being unadorned. The leading wheels of the former type were provided with outside bearings and the tenders, which were also outside framed, had four wheels. The 4–4–0 engines had six-wheel tenders with inside frames.

While Clark was at Cowlairs he had been interested in the prevention of smoke and in due course developed his own ideas and perfected his own system for effecting this desirable aim. It was not until after he had left the Great North that his apparatus was patented. The abridged specification of the patent, No. 2976 of November 30, 1857, described the apparatus as applicable to railway locomotives, marine or stationary engines. Air for primary and secondary combustion was supplied through tubular, conical, bell-mouthed or other openings above and below the firegrate, the air currents being induced by steam jets issuing from round, flat, rifled or other nozzles on steam pipes fed from the dome. These steam pipes could be arranged horizontally across the front and back or sides of the firebox and ashpan.

The first applications are recorded as having been made in England in 1858 on the North London, Eastern Counties and South Eastern Railways. However his old company, the G.N.S.R., agreed to a trial being given to Clark's invention, and a minute of the traffic committee meeting held on May 3, 1859, records that very satisfactory results both as regards consumption of smoke and saving of fuel had been obtained with an engine fitted six weeks previously. The cost of adapting it to the engines was about £10 and the royalty £10. It was resolved that all the company's engines be fitted. It was later claimed that the fuel saving amounted to around 2 lb. of coal per mile. In Manson's time the steam jets were dispensed with and the air tubes arranged back and front of the firebox instead of on the sides as originally fitted. In 1893 during Johnson's superintendence the use of these fittings was discontinued entirely.

The small companies with which the G.N.S. was associated, and with which it ultimately amalgamated, were the Banffshire, whose workings were taken over in August 1863 along with those of the Morayshire Railway, and the Deeside Railway which was leased to the Great North from September 1, 1866, and finally amalgamated with the parent company in 1876.

All these companies had their own locomotives. The Banffshire, originally called the Banff, Portsoy & Strathisla Railway, had four locomotives, two of which were 0-4-2Ts built by Hawthorn of Leith in 1859; a third, 0-4-2 goods tender, from the same works in 1861 and a Vulcan Foundry 0-4-2 tender engine bought second-hand from the Scottish Central Railway in 1860 completed the quartet. This last locomotive was one of a batch built by Tayleur & Co. in 1848. In 1868 it was purchased by Thomas Wheatley, then locomotive superintendent of the N.B.R., who combined a lucrative business in buying, repairing, and hiring out small engines suitable for contractors (and the smaller railway companies too), with his official affairs.

The fourth B.P. & S.R. locomotive was sold to the Deeside Railway and came again into G.N.S. ownership at the takeover of this line. In 1852 Neilsons built two locomotives for the Morayshire Railway. They were small 2–2–0Ts with 10 × 16 in. cylinders and 5 ft. drivers, to the designs of J. Samuel, late of the Eastern Counties Railway. Samuel was concerned with the introduction of the 'light' engine and these examples appear to be the only ones built in Scotland.

Samuel severed his connection with the Morayshire in 1854 and was succeeded by Joseph Taylor who came from the Scottish Central. Taylor added no new locomotives to the stock but repaired the existing ones and it was when testing one of these in 1857 that he met his death in a collision with another engine. Taylor was followed first by Robert Blackwood, then by George Golightly; both are described by Vallance as 'shadowy figures of whom little appears to have been recorded'. Golightly held office when the workings were taken over by the G.N.S. in 1863.

Neilson & Co. built two more locomotives for the Morayshire in 1859. These were 2–4–0s with saddle and side tanks. Rather odd and ugly in appearance, they were only slightly improved by having the safety valves enclosed in a brass casing rather like those to be seen in later years when Johnson came to the G.N.S. bringing some of his father's Derby design features with him.

The largest of these small companies was the Deeside, which had no fewer than eight locomotives. Hawthorn of Leith built the first two in 1854. They were 0–4–2T similar to, but smaller than, those built for the B.P. & S.R. five years later. Dodds & Co., Rotherham, delivered an 0–4–2 tender engine in 1857. Like the Hawthorns, it had 13 × 16 in. cylinders and 5 ft. coupled wheels, but it also had Dodds' patent wedge motion. It proved to be a most unsatisfactory machine and in 1864 it was sold to one of the Wheatleys.

Between 1856 and 1866 six 0–4–2 tender locomotives were added to the stock. Of these, five were by Hawthorn and the sixth was the Banffshire engine. The five Hawthorns were given four-wheeled tenders with one exception which had a six-wheel tender on account of the increased coal and water capacity needed for working the Royal train to Aboyne from 1859 and to Ballater from 1866.

At first, the engineer of the line, John Willett, also took charge of the locomotive department. In 1864 William B. Ferguson, the secretary and general manager, took control of both departments which he retained until 1866. It was in that year that Benjamin Conner of the Caledonian Railway was called in to make a valuation of the stock prior to the line being leased to the G.N.S

The Hawthorn engines were all typical of their builder's products at that period. Salter safety valves mounted on domes on the fireboxes, cylinders 16×22 in. and coupled wheels 5 ft. diameter. They are reputed to have been rather unsteady runners, owing to the long overhang at the leading end and, as speed of operation increased, this characteristic grew steadily worse.

The 0–4–2 purchased by the Banffshire from the Scottish Central was one of a batch built by Tayleur & Co. at the Vulcan Foundry in 1848, and it was thoroughly overhauled by the makers before the sale was effected. The cylinders were 16×18 in., the coupled wheels 4 ft. 6 in. and the carrying wheels 3 ft. 6 in.

The new locomotive superintendent of the Caledonian at the beginning of 1857 was Benjamin Conner; he served his apprenticeship to an engineer named James Gray, who had a small establishment in the East end of Glasgow. Later Conner went to Neilson & Mitchell at the Hyde Park works and was made foreman erector, eventually he became works Manager, which post he left to fill the vacancy at St Rollox.

Conner produced a number of handsome locomotives, some of which won world wide renown. Sinclair had not gone over 7 ft. 2 in. diameter for his singles, using 16×20 in. cylinders and 7 ft. drivers on his earlier express engines, and 17×22 in. cylinders and 7 ft. 2 in. wheels on the later ones built shortly before he left to join the Great Eastern Railway. The goods locomotives, also of the 'Old Crewe' pattern, were of two cylinder sizes, 16×20 in. and 17×20 in., with 5 ft. 2 in. coupled wheels. The leading wheels had outside bearings and the earlier engines of the class had raised fireboxes, the later ones having flush tops. During construction of these locomotives, the Greenock works were closed and the department newly housed in St Rollox. No. 174 of the class was the first product of the new works. The later batch had tenders with six wheels, the first on the Caledonian Railway. But Conner's first Caledonian design was the famous 2-2-2 based on the 'Old Crewe' type. There was one distinctive variation however; the large diameter of the driving wheels, 8 ft. 2 in., allowed the cylinders to be mounted horizontally whereas in the typical 'Old Crewe' locomotives they were inclined. As originally built, the cylinders were 17×24 in. but the diameter was later increased to $17\frac{1}{2}$ in. on some and $17\frac{1}{4}$ in. on others. The boilers had $192 \times \frac{7}{8}$ in. tubes giving 1,080 sq. ft. of heating surface, which together with the 89 sq. ft. of the firebox made a total of no less than 1,169 sq. ft. The grate area was 13·9 sq. ft. and the working pressure 120 lb. per sq. in. The safety valves were Salter type on the dome which was mounted on the raised firebox. Later, during rebuilding, Ramsbottom valves were placed over a flush top firebox and the dome was moved to the middle of the barrel. The large driving wheels were encased in splashers perforated like the paddleboxes of a steamship, and the leading and trailing axles had outside bearings. Of a total weight of 30 tons 13 cwt., 14 tons 11 cwt. was available for adhesion. By using Gooch's valve gear, which incorporated a fixed expansion link, i.e. fixed in so far as though the link oscillated it did not rise or fall when reversing the direction of motion of the engine, the centre line of the boiler was kept down to 6 ft. $6\frac{3}{4}$ in. from rail level. The valves had $1\frac{1}{2}$ in. laps and long travel for the special purpose of ensuring as free an exhaust as possible. This was sound practice; these were high speed locomotives and the loss of heat by radiation from, and condensation in, the exposed outside cylinders would reduce the velocity of the exhaust steam considerably.

It is significant that when these engines were modified by Drummond in 1883, he not only reset the valves but replaced the originals by short-lap valves and throughly spoiled a good free-steaming, fast-running engine. Between 1859 and 1865 twelve were built, four more coming out in 1875. One, exhibited at the International Exhibition in London in 1861, found favour with Said Pasha, Viceroy of Egypt, who purchased three from the makers, Neilson & Co.

Conner designed another single in 1864 when his 7 ft. 2 in.

variety came from St Rollox works. Much the same as the earlier larger wheel type, the new batch had 17×22 in. cylinders, boilers pressed to 120 lb. per sq. in., and 1,127 sq. ft. of heating surface. Only four were built at this time and they were all rebuilt with flush top fireboxes. The heating surface was reduced to 882 sq. ft. and the weight increased by two tons to 32 tons. In 1871 a fifth engine of the same dimensions was added. This was an experimental locomotive built by Andrew Barclay & Sons, Kilmarnock, specially to demonstrate a patent method of lifting the driving wheels clear of the rails when coasting, in order to reduce frictional wear. This was not successful, and the locomotive was soon modified to conform with the remainder. The last engine of the class to remain in service, it was broken up in 1901.

In 1860 Neilson & Co. built the first engines of a new 'improved outside cylinder mixed traffic locomotive', the '197' class. In 1860–1 Beyer Peacock & Co. constructed six more and the whole class numbered 25 engines. These were 2–4–0s with 18×24 in. cylinders, and 6 ft. 2 in. coupled wheels. The leading and tender wheels were 3 ft. 8 in. diameter. The boiler was pitched 6 ft. $5\frac{1}{2}$ in. above rail level; it was 4 ft. 1 in. diameter and the barrel was 11 ft. $2\frac{3}{4}$ in. long. The heating surfaces were:— $218 \times 1\frac{7}{8}$ in. tubes, 1,229 sq. ft., firebox 79 sq. ft., total 1,308 sq. ft. The grate area was 13·8 sq. ft. and the working pressure 120 lb. per sq. in. The dome was mounted on top of the firebox with Salter safety valves the spring balances of which passed through the embryo cab roof and secured to the boiler front.

The six-wheeled tenders had square tanks containing 1,400 gallons of water. The wheelbase was 11 ft. 6 in. and a screw hand brake was fitted. The engines weighed 34 tons.

For mineral traffic the 0–4–2 was proving popular and between 1861 and 1866 64 were built. To modern eyes they were somewhat ugly machines, their only claim to beauty being in the neat brass casing for the safety valves on the firebox. The outside cylinders were $16\frac{1}{2}\times22$ in., the coupled wheels 5 ft. 2 in. and the heating surface 969 sq. ft. Salter valves set for 120 lb. per sq. in. were mounted on the dome on the firebox top but later were substituted by Ramsbottom valves when the dome was moved to the middle of the boiler barrel. At 27 tons 12 cwt. these were comparatively light engines.

By the 1860s steam locomotives were losing the primitive appearance which had so characterized their earlier decades, and were now assuming the form which was to be at once a work of art, pleasing to the eye, and at the same time, functional in the fullest sense of the word. That they would grow in size and weight was obvious and this chronicle will endeavour to show how that growth was dictated by traffic requirements and accomplished so efficiently within the confines of the rather restricted loading gauge.

Some of the early characteristics were already disappearing by 1860 as, for example, the use of round section connecting rods, a relic no doubt of the old Manchester school of pump and mill engine practice. It was still to be found on the Highland Railway, especially on the locomotives for which Allan was responsible. As knowledge of the stresses involved was gained, there was a swing to the strong rectangular section so well known to subsequent generations of locomotive men.

In place of the double cotters, gibs and cotters became fashionable, these in turn giving way to a simple bolted strap and glut fitting. The early hexagonal brasses, awkward to fit and difficult to maintain and align, were replaced by bushes having a plain square back.

Bury's bar frames did not become popular in this country. The early designs with their snugs, wedges and bolts for securing cylinders, frame stretchers, motion plates, horns etc., gave way before the plate frame developed by Stephenson and Rastrick.

The early springs were somewhat slovenly and springmaking developed slowly. At first, springs were of short span, deep camber and composed of narrow plates. Later came the wider, thicker plate springs with long span and shallow camber. In general, the method of securing the plates was by means of a central locating bolt passing through the buckle.

Both Patrick Stirling and Alexander Allan left an indelible impression on locomotive development, north and south of the border. In Stirling's domeless boiler cult was laid the foundation of his very successful designs for the Great Northern Railway after he had left Ayrshire, and in his brother James he had an ardent supporter who followed him in office. James was, in turn, succeeded by another man trained in the Kilmarnock tradition,

Hugh Smellie who, whilst he used domes on his Maryport & Carlisle locomotives, nevertheless dispensed with them on his return to his native country. Strangely enough, James Manson, another Kilmarnock man, was content to eschew tradition and incorporate domes in all his designs for the G.N.S.R. and G. & S.W.R.

Alexander Allan's contribution was undoubtedly his very strong and rigid fore end design which is known as the 'Old Crewe' design and, although examples of it were to appear from time to time on all five of the main companies, it was on the Highland that its greatest impact was to be found. Even as late as 1892 David Jones was employing the design, though his successor, Peter Drummond, would have none of it.

There was to be a third strong influence in Scottish locomotive design, that of William Stroudley, though it was not quite so direct as in the two examples already described.

Chapter IV

THE THIRD TREND-SETTER

The ten-year period from 1860 to 1870 was one in which many amalgamations were effected all over Scotland. Some were of large companies and some were of smaller undertakings, and several of these fusions contributed to the rapid growth of the two major companies, the Caledonian and North British. The Glasgow & South Western enlarged its sphere of influence by taking over the Bridge of Weir, Maybole & Girvan, Kirkcudbright, and Castle Douglas & Dumfries Railways. The small northern lines which later became the Highland Railway were the Inverness & Aberdeen Junction, Inverness & Perth Junction and the Ross-shire Railways and the Caledonian enriched itself by absorbing the Scottish Central. The North British countered Caledonian efforts to monopolize the Edinburgh & Glasgow by absorbing it, and at the same time took over the Monkland Railways though the latter were in fact a part of the E. & G. by a matter of a few days. All these mergers took place in 1865 and in the following year the Caledonian took under its wing the Scottish North Eastern, itself an amalgam of the Scottish Midland Junction and Aberdeen Railways, thereby completing the Caledonian domination of the main trunk route from London to Aberdeen on the Scots side of the border.

On taking charge of the North British locomotive department in 1855 William Hurst was faced with what was virtually a threefold task. He had to reorganize the working of St Margaret's as regards the repair of both engines and wagons. Smith had failed to run the works with the degree of efficiency required and expected by the board with the result that locomotive casualties were occurring at an alarming rate, and the number of wagons under and awaiting repair was far exceeding the output of repaired vehicles. Smith's successor, the Hon E. G. Petre had done little to recover the situation when he faced dismissal for intoxication. During the investigation into the state of affairs towards the end of Smith's career, Hurst had been loaned from the Lancashire & Yorkshire Railway, together with their carriage and wagon superintendent, for the purpose of rendering an independent report to the board on the conditions obtaining at St Margaret's, so he was on the spot and applied for Smith's job when it became vacant. Although unsuccessful on this occasion he was selected very readily some ten months later when Petre left the service.

William Hurst was the first of the N.B.R. superintendents to produce his own designs and build his own engines at St Margaret's works. However it was not until he had been in office for five years that any of his new work materialized, the intervening years being fully occupied in clearing up the backlog of work left by his predecessors and in reorganizing the department and workshops.

The first passenger locomotives to Hurst's design were Nos. 90–5 built in 1860 by Neilson & Co. and delivered to the N.B. over E. & G. metals to the General (later Waverley) station through the Mound tunnel. These six engines were 2-4-os with 16 × 20 in. cylinders and 6 ft. coupled wheels. Together with six very similar engines built by Dübs & Co. in 1865 these were the only express locomotives the N.B.R. possessed at this period. Both classes had mixed frames, the boilers had domes mounted on the fireboxes and the Dübs engines had plain stovepipe chimneys.

Locomotives for freight work also occupied Hurst's attention and in 1860 he produced two o-6-o tender engines with 5 ft. wheels and 15 × 24 in. cylinders. The following year he brought out two more. These had $15\frac{1}{2}$ in. cylinders but were otherwise similar; then during 1862–3 Hurst purchased six complete sets of parts from Hawthorns and Stephensons and built the locomotives in St Margaret's works. The stroke of these was only 22 in.

There was nothing noteworthy about any of these designs and Hurst is perhaps best remembered for the series of very small tank engines he built for the Peebles & Hawick line and other branches. The first engine appeared in 1857 and was the first locomotive to be built at St Margaret's. Of the 2-2-2 type, it was a small well tank with 12 × 18 in. cylinders and 5 ft. 6 in. drivers. A similar engine followed and was St Margaret's second engine. These two engines bore the numbers 31 and 32 respectively. No. 31 had a small dome with a plain brass casing but No. 32 had a more ornately finished dome which was quite large and had a bell mouth and a fluted brass casing. When Thomas Wheatley fell out with the N.B.R. management and became manager of the Wigtownshire Railway he bought both these tiny engines and, what is more, he and his son rebuilt them under extremely primitive conditions in the goods shed at Wigtown for want of better or more commodious premises.

Hurst's other tank engines were all 0–4–2s with 12×18 in. cylinders and 4 ft. 9 in. coupled wheels, and were built between 1857 and 1864.

At the end of 1866 Hurst retired from the service of the N.B.R. and was succeeded by Thomas Wheatley. He had been 11 years with the company and at the latter end had been concerned with the amalgamations and found himself one of two locomotive superintendents on the staff of the new company. His colleague was Samuel Johnson of the Edinburgh & Glasgow Railway at Cowlairs, whose works manager was William Stroudley.

In order to appreciate how Stroudley's presence on a Scottish Railway was to have a marked effect on future development it is necessary to review briefly the work of the previous Cowlairs chiefs.

Paton appears to have been content to purchase his motive power from such builders as Bury and Hawthorn until 1854, when the need for more powerful engines became evident. In that year Paton obtained two Sharp singles generally similar to six very successful engines bought six years previously, but with one marked difference which was very much in their favour.

In the earlier 'Sharpies' the inside frames extended from the front buffer beam to the firebox only. This construction was not sufficiently rigid and in the later engines the inside frame was extended to the full length of the engine. This was a big improvement and credit for it must go to Sharp's designer, Charles Beyer, a man who, in addition to being a most able designer, also managed to incorporate a far greater degree of aesthetic beauty into his work than did many of his contemporaries. When Beyer left Sharp Roberts in 1855 to set up his own factory in partnership with Richard Peacock, Paton transferred his custom to the new firm.

In 1856 Beyer Peacock & Co. built six 2–2–2 express locomotives for the E. & G.R. The cylinders were 16×20 in. and the driving wheels 6 ft. 6 in. diameter; the wheelbase of 14 ft. 6 in. was equally divided between the three axles. The leading and trailing wheels were 3 ft. 6 in. diameter. Stephenson link motion controlled the valve events and well designed ports gave good admission and exhaust of steam. The steam ports were $1\frac{3}{8}$ in. $\times 13$ in. and the exhaust ports $3\frac{1}{2} \times 13$ in. The leading and trailing wheels were provided with outside bearings $4\frac{1}{4}$ in. diameter and 8 in. long, and the driving wheels had inside bearings $6\frac{1}{2}$ in. diameter by $7\frac{1}{2}$ in. length. Laminated springs were fitted above the running plates. The boiler barrel was 3 ft. $11\frac{7}{8}$ in. diameter and 10 ft. 1 in. long. 171×2 in. tubes furnished 928 sq. ft. of heating surface, the 83 sq. ft. of the firebox making a total of 1,011 sq. ft. The grate was horizontal, 14 sq. ft. in area with two sets of firebars each 2 ft. long carried in comb bars.

Salter safety valves set to blow off at 140 lb. per sq. in. were mounted on the firebox top. From the levers the lower ends of the balance springs were attached to the firebox doorplate a little above and to each side of the firehole door. Crosshead pumps supplied water to the boiler through check valves on the centre line of the barrel 3 ft. 9 in. behind the smokebox, the pumps themselves being fixed on the insides of the frames. The regulator handle was of the pull out type. These engines weighed 22 tons 15 cwt., 7 tons rested on the leaders, 10 tons 5 cwt. on the drivers and 5 tons 10 cwt. on the trailers. Two similar engines were built in 1861 arriving from Manchester in March.

The success of the Beyer Peacock singles prompted the placing of further orders for 2–4–0 and 0–4–2 types. Nos. 40 and 41 arrived in November 1859 followed two years later by Nos. 4 and 7. These were 2–4–0s with 6 ft. coupled wheels. In other respects they were identical with the 2–2–2 engines. So were the 0–4–2 goods engines which came from Beyer Peacock's works in 1859, 1861 and 1862, these engines being numbered 89–92, 93–6 and 97–100. The tenders of the coupled engines were again four-wheeled and carried 1,000 gallons of water. The makers' records show that the original six 2–2–2s had tenders weighing 7 tons 2 cwt. 1 qr. but do not show the capacity, similarly the 1,000 gallon tenders, like the earlier ones, were four-wheeled on a 9 ft. wheelbase, but the weight is not quoted.

When renumbered after the 1865 amalgamation the singles,

which had been E. & G. Nos. 6, 23, 42, 43, 56 and 57 followed by 1 and 3, became N.B.R. Nos. 211–8, the 0–4–2s became 317–328, and the 2–4–0s, 237, 238, 233 and 234. Under Johnson's management three more of these 2–4–0s were built in 1865–7 and had domeless boilers and were allotted to passenger workings. These last engines were Nos. 235, 236 and 239. All these engines were rebuilt by Drummond or Holmes and lasted until 1910–11. The Beyer Peacock engines had domeless boilers as a rule and this was a new departure for the E. & G.R. as previously its engines had had domes mounted in almost every conceivable position on the boiler or fire-box, but when they were rebuilt, two by Wheatley and the rest by Drummond, they were given domed boilers pressed to 140 lb. per sq. in.

This process of rebuilding was started by Wheatley taking No. 218 in hand in 1873. The work was finished off by Drummond and the rest of the class was dealt with by Drummond; it was not until 1912 that the last of the series disappeared from the North British scene.

Paton was interested in Clark's experiments in smoke abate-ment. Coke was an expensive fuel and the E. & G. and its neighbour, the Monklands Railways, sat on one of Scotland's richest coalfields. Paton turned to the possibility of using coal as fuel without the added expense of processing it in the coke ovens at Falkirk, which also added to the transport charges.

Thus in 1850 Paton designed an unusual locomotive. It was an 0–4–0 mineral engine with a coal burning firebox. An arrangement was applied to this locomotive, the purpose of which was to secure complete combustion of the coal. To enable this to be achieved a jet of steam was introduced into the firebox just above the fire level, the intention being that the hydrolysis of the steam would provide sufficient oxygen to complete the process of combustion. The firebox was a very long one for those days, seven feet, and it was very shallow. The boiler and tubes were short, being only about 8 ft. long. A very large dome was placed just behind the chimney. The cylinders were 15×22 in. and were secured on each side of the boiler, were steeply inclined, and drove the rear pair of coupled wheels. The experiment was not a success but it appears to have given Clark the germ of the idea from which he ultimately devised his own system of smoke prevention.

At this time Clark had been engaged in a series of experiments to determine, amongst other things, the relative merits of inside and outside cylinder locomotives. Both Paton of the E. & G. and Sinclair of the Caledonian had co-operated to the extent of allowing Clark considerable freedom to carry out his experiments, and his findings were later published in his *Railway Machinery*. Paton was unconvinced by Clark's arguments in favour of outside cylinder locomotives of the Allan or 'Old Crewe' type, and continued his policy of using inside cylinders and inside frames. He retired from the service in 1861 and was succeeded by William Steel Brown who came from the Great Northern Railway at Peterborough. Brown was in need of a works manager and his erstwhile foreman fitter at Peterborough, William Stroudley, was appointed to the post. This appointment was to prove one of considerable significance in Scottish railway history since it gave Stroudley the scope he required for the introduction and development of some of his ideas in locomotive design, and to provide a very definite and characteristic school of thought which passed down through his assistants, notably the Drummond brothers, making him a third 'trend-setter'.

Brown had not been long in command at Cowlairs before he expressed the opinion that a main line railway should not be dependent on outside contractors for the construction of its new locomotives, content only to repair and maintain them, but that it should build its own, or at least a proportion of its own, locomotive stock. Thus in 1862 a new series of 2–4–0s appeared on the E. & G. metals, built this time in the Cowlairs works under the supervision of Stroudley. There was nothing outstanding about these engines. In appearance they resembled the Hawthorn rather than the Beyer Peacock style, possibly as a result of Brown's long experience of Hawthorn locomotives on the G.N.R. These Brown-Stroudley 2–4–0s had inside cylinders 16 × 22 in. and 6 ft. coupled wheels. The boilers were domeless and the frames were double with outside cranks. The first two were built in 1862 and two more appeared in 1863. The E. & G. numbers were 101–4, later to become N.B.R. Nos. 351–4.

At a board meeting held on May 30, 1864, the death of W. S. Brown was announced and in due course Samuel Waite Johnson was appointed to the vacant position.

After serving an apprenticeship with E. B. Wilson in Leeds Johnson had joined the Great Northern Railway where his father held a position in the engineering department. As assistant loco- motive superintendent of the southern area under Sacré, Johnson was able to familiarize himself with locomotives by other builders than Wilson, and his experience was widened when he was made works manager at Peterborough. In 1859 he left the G.N.R. and spent five years acting as locomotive superintendent of the Man- chester, Sheffield & Lincolnshire Railway, remaining with this company until his appointment to the chair of the Edinburgh & Glasgow locomotive department.

Johnson first built four more of the 2–4–0 type Brown had produced in 1862. These engines bore E. & G. numbers and later became N.B.R. Nos. 355, 356, 349 and 350. It is possibly because the lowest numbers borne by this class were the last built, in 1866, that gave rise to the oft quoted statement that the 349 class was designed by Johnson when in fact Brown deserves the credit for their introduction. Between 1859 and 1861 Beyer Peacock built twelve 0–4–2 domeless-boilered locomotives with four-wheel tenders. These engines had 16×22 in. cylinders and 5 ft. coupled wheels. The boilers were 4 ft. diameter and contained 172×2 in. tubes giving 934·3 sq. ft. of heating surface and the firebox of 85·3 sq. ft. produced a total of 1,019·6 sq. ft. In many particulars they were very similar to the previous Beyer Peacock engines. The E. & G. numbers were 89–100 and they became N.B.R. 317–328. Delivery commenced in 1879 and the order was completed in 1862. Since these engines proved very successful Johnson built six more at Cowlairs in 1864 and numbered them 83–8, later N.B.R. 329–334. He also replaced an old Bury goods engine whose boiler had blown up by a double frame 0–4–2 of similar dimensions to the Beyer Peacock class.

The colliery areas served by the company abounded in sharp curves and, as on the G. & S.W.R. use was made of the flexibility of the 0–4–0 tender engines. Johnson obtained two from Hawthorns in 1864 and they became N.B.R. Nos. 268 and 269. Their use at that date emphasizes the continued avoidance of potential troubles due to too long and rigid a wheelbase, though there were signs that the 0–6–0 type would sooner or later be found necessary. In point of fact three of this wheel arrangement did appear about the time of the amalgamation, the first, No. 282 in the N.B. stock, was delivered to the E. & G. followed by Nos. 209 and 210 to the N.B.R. They were built by Dübs & Co. and had 16×22 in. cylinders, 4 ft. 2 in. wheels and double frames.

In 1865 the amalgamations of the Monklands Railways with the Edinburgh & Glasgow, and of the new combine with the North British Railway, under that title, took place, and the new board decided on an arrangement which was to prove somewhat unfortunate. It was resolved to have two locomotive superintendents. William Hurst was to remain at Edinburgh and be responsible for the original N.B. area, while the E. & G.R. under Johnson was to form the Western Division. Johnson's works manager was Stroudley and one of his young foremen was a man named Dugald Drummond who had come back to Scotland from Peto, Brassey & Betts, Canada Works, Birkenhead, whence he had gone for experience after his early training in Scotland.

All three men were of strong character and it appears that the relationships which formed between them were not by any means the most amicable. Stroudley was the first to make a move. When the Highland Railway advertised for a successor to William Barclay, Stroudley was appointed and he took the young Drummond with him to Inverness to be his foreman erector. Then in 1866 Johnson applied for, and obtained, the post of locomotive superintendent of the Great Eastern Railway in succession to Robert Sinclair, and Hurst intimated his desire to retire from the service.

The all-round withdrawal was a suitable time to effect a change and the board decided to appoint one man to run the whole department.

Their choice was Thomas Wheatley, a good sound engineer, but a man with a strange career. His initial training had been gained on the Leeds & Selby Railway after which he saw service as a driver on the Midland Railway. In 1845 he had been involved in a collision between the light engine he was driving and the rear of the Mail train to the assistance of which he was proceeding. A passenger was fatally injured: at the subsequent inquest the jury returned a verdict of manslaughter against Wheatley. However, Wheatley had not remained in court to face the result and he next turned up on the Manchester, Sheffield & Lincolnshire Railway where he was put in charge of the locomotive depot at Great Grimsby. From there he had gone to the southern division of the L. & N.W.R. at Wolverton under McConnell, which post he vacated to succeed Johnson at Cowlairs.

Wheatley was now 46 years old and he faced quite a considerable task. The heterogeneous collection of motive power units owned by the enlarged company was, with a few exceptions, no very great asset.

It is generally accepted that the E. & G. engines were in much better order than those of the eastern company, though Hurst had done much to improve conditions on the N.B.R.

Before embarking on any new work Wheatley took stock. In addition to the types already described he found a mixed bag in varying states of mechanical efficiency.

In a rash moment the E. & G. had agreed to work the Stirling & Dunfermline Railway, from which company there came a number of engines of different types, mostly by Hawthorns of Leith. The S. & D. had first been worked by the Scottish Central until the workings were taken over by the E. & G. in September 1853. Then the latter company tried to break the agreement and the little S. & D. appealed to the Railway Commissioners who decided in its favour. The larger company finally absorbed the smaller in June 1858.

In 1862 the absorption of the Glasgow, Dunbartonshire & Helensburgh Railway brought some Allan type engines into the E. & G. orbit. These were some 2–2–2s with 6 ft. drivers built in 1857–8 by Jones & Potts of Newton-le-Willows. Like some Allan type 2–4–0s also with 6 ft. wheels built at the same time by the same firm, they were slightly modified versions of the engines built for the Caledonian Railway. This amalgamation brought in the Caledonian & Dunbartonshire Junction Railway, the little Balloch to Bowling line over which Patrick Stirling had once presided. From this concern came four 2–2–2T built by Neilson & Co. in 1850. Three of them had 14×20 in. cylinders and 5 ft. driving wheels, but the fourth was similar to a sister engine named *Atalanta* on the parent company and had cylinders only 10×14 in. This diminutive locomotive later became the locomotive superintendent's inspection engine.

The amalgamation of the Monkland Railways with the E. & G.R. in 1865 brought a number of Neilson and Hawthorn engines of various kinds into the new organization. Among them were outside cylinder o· 4–2s by Neilson & Co. and Hawthorn & Co. and some o–6–os by the latter firm, and except for Paton's early banking engines, the first o–6–os on the N.B.R. All these had 16 × 24 in. cylinders and 5 ft. drivers except one engine with wheels 4 ft. 6 in. diameter. The N.B. numbers allotted to this batch were 267–279, No. 277 being the 4 ft. 6 in. engine. The rest of the Monklands stock does not merit special comment and mostly comprised o–4–o mineral engines and pugs.

At Burntisland, where the works of the Edinburgh, Perth & Dundee were situated, Robert Nicholson had put his locomotive stock in a reasonably good state of repair. His locomotives were, again, mainly from Hawthorn and Neilson. Some of the stock was originally off the Edinburgh & Northern which the E.P. & D. had absorbed in 1849. These were 5 ft. singles with 15 × 21 in. cylinders built by Hawthorn & Co. in 1846. Ten similar engines ostensibly for the same line arrived on the N.B.R. when new, apparently because the E. & N. could not pay for them. Five other singles, six footers, with 15 × 20 in. cylinders were rebuilt by Nicholson at Burntisland and later became N.B.R. Nos. 124–8. Another, No. 147, was an inside framed single which Nicholson had built to his own design. For goods work the E.P. & D. had a number of o–6–os with inside frames, 16 × 24 in. cylinders and 5 ft. wheels built in 1861, and some o–4–o mineral engines built in 1860 and useful in the Fife colliery areas.

It is evident that Wheatley was confronted by a major problem. In the great amalgamations of 1865 the child became the father of the man inasmuch as the North British Railway, once little more than the protege of the Edinburgh & Glasgow, had by absorption and amalgamation become the dominant influence, and now ruled the new and fast growing company. Nevertheless when Wheatley started his process of reorganization it was the E. & G. works at Cowlairs which were chosen as his main plant. St Margaret's was retained as a repair and maintenance depot only, though a few engines were built there during the period 1867–9. At Burntisland all new building ceased after 1862.

Two small o–4–os were acquired in 1867. They had originally been supplied to a coalmaster named Watson, who had collieries in the Cambuslang and Motherwell areas, for work around his pits. Hurst had provided the specification and Neilson & Co. built them. Of an extreme simplicity these two engines had 16×22 in. cylinders and the wheels were 5 ft. 1¾ in. diameter. The boiler barrel was 4 ft. 4½ in. diameter, 10 ft. 10 in. long and the total heating surface of these little engines was 890·7 sq. ft. The grate area was 13·35 sq. ft. and the working pressure is believed to have been 140 lb. per sq. in. The tenders, carried on four wheels, had a capacity for 3 tons of coal and 1,430 gallons of water. Holmes rebuilt both engines, one in 1885 and one in 1887. The main differences were that the boilers afterwards contained 165×1¾ in. tubes and the wheels were reduced to 5 ft. Although the tender water capacity was reduced to 1,377 gallons the total weight was increased to 47 tons 11 cwt., an increase of 3 tons 8 cwt. Several alterations were made during the Wheatley/Drummond years including the provision of stovepipe chimneys and the Roscoe lubricators were replaced by tallow cups. Later Holmes gave these engines his standard type of chimney and mounted domes on the front rings of the boilers, close behind the chimneys. That they were useful little engines is evidenced by the fact that they had long lives. One worked in the Monkland district and the other in the Edinburgh area and Leith Docks. They were withdrawn in 1909 and 1911.

The year 1868 saw the first of Wheatley's locomotives emerge from Cowlairs works. It was one of two 0–4–0 goods engines which had four-wheel tenders. Whilst the boilers, which were domed, were new, the rest of these machines had been built from scrap materials. The cylinders were 15×24 in. and were between the frames; the coupled wheels, 5 ft. 1 in. diameter, were at 7 ft. 6 in. centres. The heating surface was only 813·5 sq. ft. Originally numbered 357 and 358, they were put on the duplicate list in 1892 and again renumbered in 1895 and 1901 when they became 1010 and 1011. The former was sold for scrap in 1921, 1011 lasted into L.N.E. days and had the distinction, when scrapped in 1925, of being the last four-wheeled tender engine to work on a British main line railway.

During Hurst's term of office a start was made to rebuild some of the early singles. The first to come from St Margaret's works was No. 36, one of the 1847 Hawthorns which retained the same wheel notation. It was followed in 1867 by Nos. 37 and 38 which emerged as 2–4–0s. These engines were constructed on new frames and had domeless boilers and the coupled wheels were 6 ft. 1 in. diameter. At first the cylinders were 16×21 in. but those of No. 38 were replaced by 16×24 in. cylinders in 1868 or 1869. For a while No. 38 was employed on commuter traffic between Balloch and Glasgow and gave a good account of herself in the morning and evening races against the Caledonian engine on the adjacent line between Dumbarton and the city. In 1893 Holmes gave No. 37 new cylinders and also fitted a Stirling type of cab.

Wheatley favoured the 0–6–0 wheel arrangement for his freight locomotives and built a number of these with $16\frac{1}{4}$×24 in. cylinders and 4 ft. 6 in. or 5 ft. wheels between 1867 and 1869. His standard goods had 17×24 in. cylinders and 5 ft. wheels. The boilers were domeless and he adopted the stove-pipe chimney similar to the Caledonian design. In 1867–8 Neilson & Co. supplied twelve of these and in 1868–9 Dübs & Co. turned out fourteen. With the completion of these orders Wheatley ceased all contract building apparently satisfied that his reorganized department could cope with future requirements. This was no mean feat and was in some measure due to the ruthless policy of scrapping obsolete and uneconomical locomotives which were liabilities rather than assets. Where there was some worth in an engine it was rebuilt and in the course of his eight years at Cowlairs some seventy were so treated.

When William Stroudley arrived at Lochgorm Works in Inverness in 1865 he was 32 years old and he had had a very good grounding in the principles of the profession he had chosen to follow. After an apprenticeship in Birmingham under a Tubemaker and Enginewright, he spent a year at Swindon under the tutelage of Daniel Gooch. Thus early in Scottish locomotive history does the Swindon influence intrude, though without any pronounced effect at this stage, since Stroudley had ideas of his own as was to become apparent as his career developed. He had shown his mettle while serving on the G.N.R. at Peterborough.

At this time, locomotive working on 'The Hill', as the Perth-Inverness road was known, was a little involved. From the opening of the Forres-Aviemore section in 1863 the locomotive power was provided by the I. & A.J.R. who also found the coaching stock, but the Perth & Dunkeld Railway, opened on April 7, 1856, had been

12. Edinburgh & Glasgow Railway, 2–2–2 No. 23, Beyer Peacock. Built May 1856. Order No. 66. Works No. 22. Photographed outside Garton Works. Cyls. 16 x 20. D.W. 6 ft L & TW 3 ft 6 in. *By permission of Manchester University.*

13. North British Railway, 0–6–0 No. 67, Built Hawthorns, Leith, No. 595, 1850. Rebuilt St. Margarets, July 1869. Renumbered 67A, 1882; 484, 1895. *Railway Magazine.*

14. Highland Railway, Stroudley's snow plough, largest size, attached to the leading engine of four 2–4–0.
British Railways.

15. Scottish North Eastern Railway, 2–2–2 No. 461, Vulcan Foundry, 1865.
English Electric Co.

operated by the Scottish Midland Junction until the end of May 1863, by which time the s.m.j. had been amalgamated with the Scottish North Eastern. The opening of the line from Dunkeld to Pitlochry on June 1, 1863, was followed by the opening of the Pitlochry-Aviemore section on September 9th of the same year providing at long last a more direct route between the Highland capital and the south than the roundabout route via Keith and Aberdeen.

At first the s.n.e. worked the line from Stanley Junction to Pitlochry but this company's resources were insufficient to meet these needs for long so the Scottish Central took over the locomotive duties from the beginning of August that year, the i. & p.j. supplying the vehicles. As soon as the line was open throughout, the Perth & Dunkeld was absorbed by the i. & p.j. and the negotiations which culminated in the amalgamation of the latter company with the i. & a.j. to form the Highland Railway came to fruition by the Act of Parliament of February 1, 1865.

The financial instability of the railway industry at this time meant that there was no money to spare for the construction of new engines. Consequently it was not until 1869 that Stroudley turned out a new design. However, he was by no means inactive and left a decided mark on the Highland Railway, directly by his own contributions and indirectly by his association with the Drummond brothers; Dugald, who was his foreman erector at Lochgorm and followed him to Brighton, and Peter who joined him at the latter place and returned to Inverness in 1896.

Stroudley was concerned to improve the standard of locomotive performance on 'The Hill' and to this he directed his attention. The engines in use on passenger trains were sadly lacking in adhesive power and the negotiation of the severe gradients was a sore trial to them, to the enginemen and to the passengers alike, to say nothing of the board who received the reports and complaints of bad running. By far the better performers were the 2–4–0 goods locomotives, so Stroudley took the sensible course of rebuilding No. 1, *Raigmore*, the first of the Hawthorn 2–2–2 of 1855, and in her rebuilt state incorporated some interesting features.

The cylinders were bored out to $15\frac{1}{2}$ in. diameter, and by using eccentric crank pins, the stroke was increased to 21 in. This required the cylinder covers to be reduced in thickness and the employment of thinner pistons. The result was a useful and businesslike engine with four coupled wheels instead of a single pair of drivers which therefore improved adhesion and gave an increase in tractive power of 12½ per cent. Moreover the engine was provided with a cab for the protection of the enginemen. Barclay, indeed, had produced an embryo cab intended to give some measure of comfort and shelter from the extremely rigorous weather conditions to be met with on the more exposed parts of the line during the highland winters. This was not Stroudley's first attempt at this detail of design; whilst at Cowlairs he had obtained the authority of his chief and had fitted a cab to an Edinburgh & Glasgow 0-4-2 No. 85, later N.B. No. 331, one of the 1859 engines.

Alexander Allan on the Scottish Central Railway had fitted primitive cabs to some of his engines and nephew Barclay had improved on them by adding side sheets on some of the Hawthorn singles and 2-4-os turned out in 1862 for the Inverness & Ross-shire Railway, the nucleus of the line to the far north, authorized in 1860 and opened in 1862. The Ross-shire line started from a junction with the Inverness harbour branch, crossed the Caledonian Canal by means of a swing bridge and ran to Invergordon on the Cromarty Firth. From the first it was worked by the I. & A.J.R., with which company it was amalgamated in 1862. Another point of interest attaching to the Ross-shire engines built by Hawthorn & Co. is that they were equipped for coal burning by having Beattie's transverse mid-feather in the firebox.

In the year 1866 Stroudley invented and patented a design of facing points with a locking mechanism which had the approval of the directors and was favourably demonstrated to the Board of Trade representatives on the directors' recommendation.

The severe highland winters with their accompaniment of snowstorms and blizzards set the engineers a problem which was resolved in a workmanlike manner by Stroudley and his colleague John Buttle, the permanent way superintendent. Stroudley designed and constructed in Lochgorm works three sizes of snow plough which could be fitted to the front end of a locomotive. The smallest plough was of light construction and it could be fitted to all engines running on the line: it was capable of clearing snow to a depth of from 12 to 24 inches. A heavier type of plough which could cope with snow up to a depth of 5 feet was designed to be attached to a pilot engine, whilst the third and largest type of plough, capable of charging drifts up to 10 or 11 feet deep, required four or even five engines to propel it. So effective were these measures during the winter of 1866–7 that they were given special mention in a paper read by the chief engineer, Joseph Mitchell, before the British Association in Dundee the following September. While the country suffered from severe snow blocks in other places the Highland line remained clear by the use of Stroudley's ploughs.

In the same year, 1869, that the rebuilt *Raigmore* emerged from the works, Stroudley built his one and only Highland engine. The new engine was an 0–6–0 tank engine whose boiler came out of No. 3, *St Martins*, one of the 1856 Hawthorn singles. Shortened to a length of 7 ft. 9½ in., the boiler now had a heating surface of 671¼ sq. ft. made up of: tubes 608 sq. ft., and firebox 63¼ sq. ft., but retained the original 12¼ sq. ft. grate. The working pressure was raised to 120 lb. per sq. in. and the cylinders were 14 × 20 in. and the wheels 3 ft. 7 in. diameter. Stephenson motion was fitted to this engine but the Allan gear continued to be used and was strongly favoured for many years. Stroudley's successor, Jones, built two more similar engines in 1872 and 1874. As was customary on the Highland all bore names appropriate to the area, these three being No. 56 *Balmain*, No. 57 *Lochgorm* and No. 16 *St Martins*, the second engine of that name. All were subsequently rebuilt in 1895–7 when the heating surfaces were modified slightly and the wheel diameter increased to 3 ft. 8 in.

The side tanks of these engines were arched over the boiler and it is generally considered that the class formed the prototype for Stroudley's L.B. & S.C.R. 'Terriers', produced in large numbers when he had established himself at Brighton, where he went in 1869. His resignation was handed to the Highland board and duly minuted on December 7 of that year.

During the summer of 1865, a very busy period of parliamentary business in relation to Scottish railways, the Caledonian acquired the Scottish Central, which had itself absorbed several smaller companies.

Benjamin Conner, the locomotive superintendent of the Caledonian, now found himself responsible for a large number of locomotives in addition to his own stud of Caledonian engines. In general, the influence of Alexander Allan predominated though there were some other designs as well. Both the Scottish Central and the Scottish Midland Junction had a number of 'Old Crewe' type singles, mostly built by Jones & Potts and there were some by Tayleur & Co. and by Scott & Sinclair as well as others built in the Greenock works of the Caledonian. There were also some single framed 0–4–2 engines with underhung outside cylinders.

The s.m.j. and Aberdeen Railways which, by amalgamation in 1856, had become the Scottish North Eastern had a number of engines of various types, including the two Cramptons. Most of the locomotives taken over by the Caledonian were not very serviceable and quickly found their way to the scrap heaps. There was however, a quite useful series of twenty-four 0–4–2s with outside cylinders 17×24 in. and 5 ft. coupled wheels. These had been designed by Thomas Yarrow and had 1,073 sq. ft. of heating surface, 14 sq. ft. of grate and a working pressure of 120 lb. per sq. in. Stephenson & Co. delivered a tank engine to the Arbroath & Forfar Railway in 1864; a smaller edition of the Yarrow engines, it was required mainly for passenger trains. In 1865–6 eight 2–2–2 engines were supplied by Vulcan Foundry and a similar one was built at Arbroath. These had 16×20 in. cylinders and driving wheels 7 ft. 1 in. diameter.

Perth became the main shop for locomotives on this more northern section of the Caledonian system and in 1868 two engines were turned out from these works. The design had been initiated by Yarrow and work on them commenced at Arbroath. They were taken to Perth where they were completed and painted in c.r. style.

Once again the Allan type of fore end was in evidence. The cylinders were 17×22 in., coupled wheels 7 ft. 2 in. and the leading wheels 3 ft. 10½ in. The coupled wheelbase was 8 ft. and the total wheel base 15 ft. 6 in. The boiler barrel was 4 ft. 1 in. diameter and 10 ft. 11½ in. long, containing 227 tubes 1¾ in. diameter, providing a heating surface of 1,040·61 sq. ft. To the 72·81 sq. ft. of firebox surface was added 28·61 sq. ft. offered by the mid-feather giving a total of 1,142·03 sq. ft. The grate area was 13·52 sq. ft. and the working pressure 120 lb. per sq. in. Two safety valves were fitted, a pillar type on the boiler barrel and a Salter type on the dome mounted on the firebox. To avoid the use of separate balance weights the rims of the driving wheels were $\frac{5}{8}$ in. eccentric so that the thickness of tyre and rim opposite to the big ends was $2\frac{1}{8}$ in. and the minimum, on the arc containing the big ends, $\frac{7}{8}$ in. The trailing wheels were concentric and compensating beams connected the laminated springs of both axles. Underhung laminated springs were provided for the leading axle which had outside bearings, those of the coupled wheels being inside the frames. The tenders of these two engines were four-wheeled and carried 1,800 gallons of water. A distinctive feature was the cab, typically Stroudley in design, and one of the earliest in the country. These two engines were numbered 472 and 473 in their new owner's stock but were renumbered 123-4 in 1876. Generally employed on passenger work, No. 123 between Perth and Forfar and 124 to Crieff they lasted, with various overhauls, until 1888 and 1893 respectively.

A class which became well known, the last example surviving until 1913, was Conner's No. 1 class 2-4-0. The first of these appeared in 1869 and in the next two years the class was built up to a total of twenty-nine. Both Dübs and Co. and Neilson & Co. supplied these engines, but the products of each builder varied slightly. They could be identified by the safety valves; Neilson's engines had spring balances in a brass casing while the Dübs engines had either Naylor or Salter valves. The cylinder dimensions were $16\frac{1}{2} \times 22$ in. and the coupled wheels 6 ft. 2 in. with leading wheels 3 ft. 8 in. diameter. During the ensuing decade the cylinders were enlarged to 17 in. The heating surface of 982·2 sq. ft. was made up of 914 sq. ft. of tube surface and 68·2 sq. ft. of firebox, while the grate area was 14·3 sq. ft. and the working pressure 140 lb. per sq. in. The arrangement of cylinders and framing was the customary Allan design and the springing was underhung and provided with compensating beams. The engines of both makers had the same coupled wheelbase, 8 ft. 7 in., but the Dübs batch were 1 in. longer in total wheelbase than the Neilsons, 15 ft. 2 in. instead of 15 ft. 1 in. The engines weighed 34 tons 19 cwt. 3 qr., and the four-wheeled tenders carried 4 tons of coal and 1,500 gallons of water and weighed 19 tons 5 cwt. 1 qr. When built they were used on Glasgow-Edinburgh and coast services: latterly they were allotted to station pilot duties at some of the larger stations and several found themselves on the Portpatrick Railway.

Another order met by the combined efforts of Neilson & Co. and the company's St Rollox works was for 16 more 2-4-0s with 17×24 in. cylinders, 7 ft. coupled wheels, 920 sq. ft. of heating surface and a working pressure of 140 lb. per sq. in. These were all built in 1867-8 and were to augment the work of the big 8-ft. singles. Conner's successor rebuilt those built at St Rollox; the Neilsons were similarly dealt with by Lambie twenty years later.

The 2-4-0 type was proving a useful and indeed a popular machine. In all over 200 were built for the Caledonian during Conner's time. In addition to those already described, there were thirty-one built in 1872-3 for goods work. A number of these eventually received Drummond type boilers from the 418 class rebuilds.

A further series of twenty 2-4-0s for secondary work on the main line followed. They had 17×22 in. cylinders, 6 ft. 2 in. coupled wheels and 1,071 sq. ft. of heating surface. A grate 11·7 sq. ft. in area seems somewhat small for these engines considering the work they were to perform on the main line. Nevertheless they lasted until the First World War.

Neilson & Co. and Dübs & Co. shared an order for thirty engines similar to the passenger type and turned them out in 1874-8. These too lasted until 1905-19. All these engines embodied the characteristic Allan front end design.

An attempt was made at this time to introduce a form of steam reverser. Eleven of the Neilson and seven of the Dübs engines of the 1872-4 batches were so fitted, but lubrication troubles due to hardening of the valve leathers from the lack of oil caused the early substitution of the more usual lever type of reverser.

No. 33 of this class was the first engine to have the Steel-McInnes air brake, and one of the St Rollox engines the first in Britain to be fitted with the Westinghouse brake.

Loadings of freight and mineral trains were increased and there was a need for stronger locomotives to handle these classes of traffic.

In 1874 St Rollox turned out the first five of a powerful class of 0-6-0s. These were long-boilered Stephenson type engines with all wheels in front of the firebox. The cylinders were horizontal 18×24 in. and the wheels 5 ft. 2 in. diameter. The crossheads were of the single bar type. The boilers were 4 ft. 2 in. diameter and 14 ft. 1½ in. long, made up of four rings butt-jointed; 134×2 in. tubes gave 991 sq. ft. of heating surface and the firebox provided 92·5 sq. ft. The grate area was 14·3 sq. ft. and the working pressure 140 lb. per sq. in. Some of the tenders were four-wheeled and carried 1,840 gallons of water and 3½ tons of coal. A noteworthy feature was a specially designed centre buffer in addition to the side buffers between the engine and tender: this arrangement materially assisted in the negotiation of sharp curves in such places as colliery sidings, with the result that derailments were very infrequent. This facility for taking severe curves was an asset which ensured a reasonable degree of success for the class, which eventually numbered thirty-nine engines, since their use on the main line was generally precluded by their instability, a general characteristic of this particular arrangement of wheels and boiler. It was one of the very few outside cylinder classes of 0–6–0 in the country.

Because there were so many ramifications of the Caledonian Railway into all sorts of odd corners, the four-wheeled tender was a particular favourite. It was usually provided with handrails and footboards on each side for the use of shunters. Tank locomotives very often had no coal space and the fuel had to be carried in a small four-wheeled truck of about 1¼–1½ tons capacity. Of shunting 'pugs' as tank engines were known throughout Scotland, there were twenty-three 0–6–0s and a number of 0–4–0 engines. An interesting development in 1873 was the production of four 0–4–4 well tanks for the Edinburgh suburban services. These had 17 × 22 in. cylinders, 4 ft. 8 in. coupled wheels and 2 ft. 8 in. bogie wheels. The coupled wheelbase was 5 ft. 9 in. and the bogie wheelbase 5 ft., while the total wheelbase was 20 ft. 9 in. The coupled wheels carried 26 tons 2 qr. and the bogie 16 tons 3 cwt. The cylinders were horizontal, the crossheads again being of the single bar type and the coupling rod ends and big ends were of the cottered split brass type favoured in so many of Conner's designs.

In 1876 Benjamin Conner died and was succeeded by his chief assistant George Brittain: thus a 20-year era in Caledonian locomotive history came to an end, and with the advent of Brittain a new era was ushered in.

Chapter V

A NEW STAR IN THE FIRMAMENT

The departure of Stroudley from Inverness on his appointment as locomotive superintendent of the London, Brighton & South Coast Railway was an event which was to have a far-reaching effect on the design trends of Scotland's railways.

Shortly after he settled in at Brighton he made Dugald Drummond his works manager; Dugald's younger brother Peter also went to Brighton on the completion of his early training in Glasgow.

In November 1874 the North British Railway advertised for a locomotive superintendent due to the enforced resignation of Wheatley and on December 30th Dugald Drummond was appointed at a salary of £700 per annum to take effect from February 1, 1875.

The stage in Wheatley's career reached in the previous chapter was that at which he could consider that the reorganization of his department was reasonably complete and working smoothly. He was on the verge of completing the locomotive for which he is perhaps best remembered, the four coupled express engine with leading bogie, inside cylinders and inside frames. Two were built in 1871 and, apart from the fact that one had an adventurous career, they set a pattern for years to come, the inside cylinder, inside frame 4–4–0 being found on every British main line railway except the Great Western.

These two engines, Nos. 224 and 264, had 17 × 24 in. cylinders, coupled wheels 6 ft. 6 in. diameter and weighed 38 tons, of which 24 tons 9 cwt. were available for adhesion. Laminated springs were used under each coupled wheel axlebox and the decision to use a four-wheeled bogie was determined by the sinuous nature of the N.B.R. main line and the sharp curves which abounded. A dome was mounted on the firebox and a half-hearted attempt at a cab was provided for the enginemen. The severity of sheer simplicity was avoided by the slotting of the coupled wheel splashers, other-wise the keynote would have been austere indeed, accentuated by the solid bogie wheels which were 2 ft. 9 in. diameter. The heating surfaces were made up of 208 tubes 1⅝ in. diameter, 894 sq. ft., firebox 87 sq. ft., total 981 sq. ft. and the grate area was 15·75 sq. ft., thus short enough to be accommodated between the coupled wheel axles without increasing the coupled wheelbase beyond a modest 7 ft. 7 in. Both engines were at work until 1919 and were then sold for scrap. No. 224 went down with the old Tay Bridge during the storm on December 28, 1879, and was subsequently raised, repaired and returned to service. It later participated in an interesting experiment in compounding during Holmes's superin-tendency. From the Tay Bridge misadventure No. 224 earned the sobriquet 'The Diver'. Six years later, in 1885, Holmes rebuilt No. 224 as a compound, on the tandem principle, but in 1887 it was restored to simple expansion. In the course of its long career this engine was reboilered twice, once by Drummond and once by Holmes.

The two early 4–4–0s followed a year after a pair of very good 2–4–0s with 16 × 24 in. cylinders, 6 ft. 6 in. coupled wheels, boilers with safety valves and manhole mounting over the firebox, slotted driving splashers and six-wheel tenders of 1,650 gallons capacity. So good were they that when given an extra inch on the diameter of the cylinders their work was not impaired. Numbered 141 and 164, they were relegated to the duplicate list in 1912 and scrapped in 1923. Wheatley again produced some 2–4–0s in 1873 when eight were turned out of Cowlairs works. Dimensionally very similar to the 1870 engines they underwent rebuilding in 1890 and again in 1915. The earlier engines had domes on the fireboxes whereas the latter carried domes at the centre of the boiler barrels. Six of these engines lasted into the L.N.E. era.

At this period the Midland Railway was pushing northwards towards the Scottish border, its general manager, James Allport, intent on expanding his already extensive empire by the addition of through traffic from London, St Pancras, to Carlisle and thence to Edinburgh over the mountainous Border Union line. Accord-ingly Wheatley designed and built at Cowlairs an improved 4–4–0 of generally similar appearance to the 1871 type. Four came out in 1873 bearing the numbers 420–3. The cylinders were 17 × 24 in.

and the coupled wheels 6 ft. 6 in. diameter. Again solid bogie wheels were used and the engines scaled $37\frac{3}{4}$ tons. No. 421 was later fitted with the Westinghouse brake, being the first North British locomotive to be so fitted.

The standard goods engines which Wheatley produced in some numbers were a successful class of 0-6-0, and even as late as 1922 there were thirty-seven still in service to be taken into L.N.E. stock. To meet immediate needs while the department was being reorganized Neilson & Co. built twelve and Dübs & Co. fourteen, the first engines appearing in 1867. By 1869 Cowlairs was in production and a further 62 were built there. Wheatley favoured inside cylinders and inside frames and his 0-6-0s followed this style. The coupled wheels were 5 ft. diameter and the cylinders were 17 × 24 in. A simple, almost chaste, design with no frills. Sanding apparatus of the trickle pattern was provided to the front of the leading wheels and the back of the driving wheels and the balance weights were opposite the big ends and in the driving wheels only, no proportion of the counter-balance being put in the leaders or trailers.

Wheatley did not provide much protection for his enginemen, the front weatherboard being bent over at the top and given very narrow side sheets to afford only meagre shelter from the elements. Although the Manuel smash of January 27, 1874, was primarily due to misuse of the Time Interval rules for traffic operation, the fact that tender-first running was involved, apparently needlessly, meant Wheatley had some awkward questions to answer at the Board of Trade enquiry.

1872 saw the production of two of the smallest tank engines built at Cowlairs. They were specially for shunting at Leith Docks where there were many sharp curves, thus the wheelbase was only 4 ft. 6 in. The four coupled wheels were 3 ft. 6 in. diameter and the cylinders 11 × 18 in. They were the only Wheatley engines to have outside cylinders.

The year 1874 saw six 0-6-0 saddle tank shunting 'pugs' come from Cowlairs works, mostly with 13 × 18 in. cylinders and 3 ft. wheels.

A number of the older engines were rebuilt by Wheatley including the '90' class 2-4-0 of 1860 and some of the 0-6-0 goods engines. In 1871 the N.B.R. took over the working of the Forth &

Clyde Junction Railway and took into stock from that company four Allan type 2–4–0s with 16 × 22 in. cylinders and 5 ft. coupled wheels. These engines had been built by Peto, Brassey & Betts at their Canada Works in Birkenhead, and had raised fireboxes, safety valves on dome and boiler, the dome being on the fireboxes. One, No. 404, was rebuilt by Wheatley and ultimately became the only Allan type left on the N.B.R.

Thomas Wheatley undoubtedly had a flair for organization and during his eight years at Cowlairs he had increased the annual output of new locomotives from six to forty. In addition the works had undertaken the heavy repair of a stock of some 450 loco-motives, but all was not well in the department. Wheatley lived in a company house, a villa at Lenzie, and it was alleged that he had furnished and decorated it at the company's expense. Moreover there were irregularities in the works involving a matter of the theft of some castings. The situation became untenable and his resignation was sought and obtained. Thus the way was clear for an event of great consequence in Scottish railway history, the arrival at Cowlairs of Dugald Drummond, as recounted in the opening paragraphs of this chapter.

Dugald Drummond was no stranger to Cowlairs. He had one big advantage over his predecessor: the financial depression of the middle sixties was over, and the troubles which had culminated in the failure of the banking house of Overend Gurney and the consequent run on the Bank of England, when so many investors, large and small, were broken, were things of the past. More money was available for the purchase of machines and machine-tools and for locomotives to be bought from private builders until such time as Cowlairs could produce as Drummond wanted. So for the second time in a dozen years Cowlairs underwent reorganization.

At this time the N.B.R. was increasing the number of branch lines and suburban services on its system, and there was a decided shortage of tank engines for such workings. Drummond's first contribution to the N.B. stock was therefore a tank engine modelled on the lines of Stroudley's L.B. & S.C.R. 'Terrier' type. The N.B.R. engine was slightly larger having cylinders 15 × 22 in., six coupled wheels 4 ft. 6 in. diameter, no carrying wheels, 701 sq. ft. of heat-ing surface and a working pressure of 140 lb. per sq. in. Altogether a neat, compact and workmanlike job for a first design, it was no slavish copy of the master's engines; Drummond introduced several features of his own, for example, he put Ramsbottom safety valves on top of the dome, he developed his own shape of chimney and a modification of Stroudley's cab in a somewhat simpler, but equally neat and efficient, form. In 1875 four were built, followed by two in 1876, by seven in 1877 and in 1878 the class of 25 was completed by twelve more. All bore the names of the localities in which they worked, a practice Drummond brought with him from Brighton.

The following year, 1876, witnessed the introduction of two more classes, an 0-6-0 goods engine and Drummond's first express design, a 2-2-2. The goods engines numbered 32 of which 12 were built at Cowlairs and 6 by Neilson & Co. in 1876 followed by 14 from Hyde Park in 1877.

Reminiscent of Stroudley's L.B. & S.C.R. 0-6-0 in outline and some of the detail, they had Drummond's own modifications as in the case of the previous year's tank engines. Also, the cylinders were 18 × 26 in. and the working pressure 150 lb. per sq. in. These were useful engines on the steeply graded road to Carlisle. The engines weighed 39 tons 15 cwt. and the tenders, carrying 5 tons of coal and 2,500 gallons of water, weighed 32 tons. Wooden brake blocks were used throughout, those on the engine being in front of the wheels, and on the tender behind the wheels.

When he left Cowlairs Stroudley took with him some deep impressions. One result was that his 'Grosvenor' singles of 1874 derived in part from Paton's E. & G.R. engines. It is therefore not strange that the first of Drummond's express designs should be based on the Master's engine. The two singles built in 1875 by Neilson & Co. were almost pure Brighton. Nos. 474 and 475, given the names *Glasgow* and *Berwick*, were intended for the main line between these places. The driving wheels were 7 ft. diameter and the carrying wheels 4 ft. 6 in. The inside cylinders were 17 × 24 in. The boiler was of the large diameter of 4 ft. 4 in. and contributed 1,133 sq. ft. of heating surface to the total of 1,225 sq. ft., the firebox contributing 93 sq. ft. The grate area was 16·5 sq. ft. and the working pressure 140 lb. per sq. in. Of the total weight of 37 tons 18 cwt., 15 tons 14 cwt. rested on the drivers. The wheelbase was equally divided: 7 ft. 9 in. leaders to drivers and 7 ft. 9 in. drivers to trailers. Domes were mounted on the back rings of the boilers and the safety valves, Ramsbottom pattern, were on the domes Drummond fashion. These two engines put up some creditable running between Edinburgh and Glasgow with the through carriages from King's Cross brought north by the 10.00 a.m. and 10.35 a.m. expresses. The time allowed for the 47¼ miles was 70 minutes, with two stops.

Early in 1876 the Midland Railway's Settle–Carlisle line was nearing completion, and the North British was particularly interested in securing whatever traffic it could at the border station. In its endeavours to reach the North Eastern stronghold at Newcastle, the N.B.R. had in 1862 been forced to concede running powers to the N.E.R. between Berwick and Edinburgh as the price for access to Newcastle, and the pickings from the East Coast traffic were not exactly riches. What little was left to the N.B.R. at Carlisle after the L. & N.W.–Caledonian combination had taken the choicest fruits was hardly worth having. Thus, with the advent of the newcomer to the western border town and the possibility of a richer harvest, negotiations were opened with the Midland. On March 31, 1876, Messrs Allport, Wainwright and Walker, the general managers of the Midland, G. & S.W. and N.B. Railways respectively, met in London and the proposals for through services between London and Edinburgh and Glasgow were formulated. Approval, so far as the N.B.R. was concerned, was given at a Traffic Committee meeting on April 13, 1876, with May 1st set as the commencing date. On the same date the Loco. & Stores Committee authorized the fitting of four passenger engines with Westinghouse brakes equipment at a cost of £90 per engine.

In May of the same year, as a result of experience gained during the first three weeks of the through services; Drummond recommended the construction of six powerful passenger engines to work the Midland trains between Edinburgh and Carlisle, as the Wheatley '420' class engines were unable to lift the heavy trains over the summits of Falahill and Whitrope without the aid of pilot engines. Arrangements were made for the substitution of four bogie passenger engines and tenders for four of the goods engines on order. The increased cost was £185 per engine making a total cost of £3,045 each. In designing these locomotives Drummond sought to obtain the flexibility needed on the Carlisle line which had not initially been laid out as a main line. He abandoned, once

93

and for all, the use of front-coupled wheels so beloved by Stroudley for express work. Instead he followed Wheatley's lead and adopted a four-wheel bogie, but of the Adams type instead of the short wheelbase fixed centre type Wheatley had used. The new engines were delivered late in 1876 bearing the numbers 476–9 and were named in the customary fashion. In December 1876 Drummond had staged his own version of the Newark Brake Trials, pitting the Westinghouse Automatic against the Smith non-automatic vacuum brake. As a result of these trials the North British board decided to adopt the former and in consequence the four new 4–4–0s were sent back to Hyde Park works to have the Westinghouse equipment fitted and several modifications carried out. Even so the Westinghouse fittings were on the tenders only, the engines retaining the steam brakes until a later date when the air brake was fitted throughout.

These 4–4–0s, of which eight more were built in 1878, were to form the basis of development of the type on several railways for many years, were first rate performers from the start. The cylinders were 18 × 26 in. and the coupled wheels 6 ft. 6 in. diameter. The coupled wheelbase was 9 ft., allowing a long sloping grate of 21 sq. ft. area. The heating surface was made up of 222 tubes 1¾ in. diameter giving 1,005·3 sq. ft. (though not mathematically correct for the length between tubeplates shown on the drawing, this figure is the official one), firebox 94 sq. ft., total 1,099·3 sq. ft., and the pressure was 150 lb. per sq. in. The weight of the engine was 44 tons 5 cwt. of which 30 tons 3 cwt. was available for adhesion. The tender weighed 32 tons and carried 4 tons of coal and 2,500 gallons of water, the total weight of engine and tender being 76 tons 5 cwt.

On the Anglo–Scottish services these engines were allowed 2 hr. 20 min. for the 98¼ miles from Edinburgh to Carlisle, with the two summits of Falahill and Whitrope to be climbed. The loadings were comparatively heavy including Pullman day or sleeping coaches and Midland twelve-wheelers of the 1875 type.

So far as he had gone Drummond had won success with each type of engine had had brought out, but he did have one failure, one instance where whilst following the Brighton school of thought he was unsuccessful. This was with a small 0–4–2 tank engine for use on the north bank of the Clyde, replacing the obsolete machines taken over with the Glasgow, Dumbarton & Helensburgh Railway.

These little 0–4–2T engines were kin to the Brighton 'D' class differing, however, in some dimensions. The cylinders were 17×24 in., coupled wheels 5 ft. 9 in. diameter and the trailing wheels 4 ft. 6 in. The heating surface was 1,075 sq. ft. and the working pressure 140 lb. per sq. in. With 17 tons of the total 45½ tons resting on the trailing axle it is small wonder that they were soon in trouble due to bad riding and a tendency to derailment. Fortunately only six were built, all at Cowlairs in 1877, and by 1880 all had been rebuilt. In carrying out this exercise Drummond substituted an Adams bogie for the trailing axle, lengthened the frames, and increased the bunker capacity to accommodate it. When returned to service these engines, now 0–4–4Ts, were sent to Fife depots for coast trains.

The need for replacements for the Helensburgh services led Drummond to produce his next type; this time a reversal of wheel arrangement in that he now had built by Neilson & Co. three 4–4–0 tank engines. The services on this line were notable and have already been commented on, and the new engines, Nos. 494–6, gave a good account of themselves when they appeared in 1879. The bogie wheels were the same size as those of the 4–4–0 tender engines, i.e. 3 ft. 6 in., the coupled wheels 6 ft. and the cylinders 17×26 in. The heating surface was 1,174 sq. ft. made up of 220×1¾ in. tubes 1,074 sq. ft., 100 sq. ft. of firebox area, whilst the grate was 17·5 sq. ft. and they weighed 46 tons 16 cwt.

The success of these handsome machines resulted in a slightly smaller version of the same type for local branch traffic appearing in 1880. Six were built at Cowlairs with 16×22 in. cylinders, 5 ft. coupled wheels and 701 sq. ft. of heating surface. The bogie wheels of this, the '72' class, were 3 ft. diameter and, like Wheatley's bogie wheels, were solid. In 1881 six more were built, followed by eight in 1882, four in 1883, and six in 1884, the last ten after Drummond had left Cowlairs and gone to St Rollox. These engines were a bit lighter than the three larger ones and weighed 35 tons 4 cwt.

In addition to the large 4–4–0T, 1889 saw Drummond's second 0–6–0 type which appeared in batches each year until 1883 when the last eighteen were built. Dübs & Co. contributed five in 1879 and the class totalled 105 engines. In later years Holmes built additions to the class with slight modifications. As built they had

17×24 in. cylinders and 5 ft. wheels and were typical of Drummond's own style which was rapidly developing.

In 1877 the absorption of the Leven & East of Fife Railway had brought in five small four-wheelers, outside cylindered tank engines, three of which were built by Hawthorn of Leith in 1857, and two by Black Hawthorn & Co. of Gateshead in 1874. Two Neilson 'pugs' were acquired in 1882. These popular little machines formed the basis of the dock 'pugs' used on various parts of the system. 1879 also saw the acquisition of four powerful 0-6-0 side tank locomotives which had been built by Dübs & Co. to the designs of A. Simpson C.E. engineer of the Glasgow, Bothwell, Hamilton & Coatbridge Railway two years before. These engines became N.B.R. Nos. 502-5 on delivery, the working of the G.B.H. & C.R. being in the hands of the N.B.R. representing a Joint Committee of Management. The cylinders were 18×24 in., wheels 4 ft. 6 in. diameter and the working pressure 130 lb. per sq. in. Thus the tractive effort of these useful machines was 15,900 lb. When in 1895 and 1897 Holmes reboilered them he raised the pressure by 10 lb. per sq. in. which had the effect of increasing the power by 1,224 lb. to over 17,120 lb. As built, the boilers had 1,065 sq. ft. of tube surface and 95 sq. ft. of firebox and, after reboilering 202 tubes $1\frac{3}{4}$ in. diameter gave 1,020 sq. ft. in a total 1,113 sq. ft. The grate remained at the original 17 sq. ft. The side tanks carried 820 gallons of water and the bunkers held 30 cwt. of coal.

The manner in which Drummond's locomotives were handling the traffic of the ever growing North British system, and the extent by which they and their designer had contributed to the financial improvement of the company, did not escape the notice of the Caledonian board which, when Brittain resigned in July 1882, made Drummond his successor.

During the seven and a half years that he had served the N.B.R. Drummond had added 151 locomotives to stock, either contract or Cowlairs built. In the same period his ruthless scrapping of obsolete machines had gone far to modernize the motive power fleet. Where there was genuine worth in an engine he would rebuild it and this policy was followed with a number of E. & G. engines which were in somewhat better mechanical order than those of other of the N.B. subsidiaries when taken over. In 1880 four of the Beyer Peacock 6 ft. 6 in. singles of 1856–61 were rebuilt

16. Caledonian Railway, 2–4–0 No. 214, built St. Rollox, 1862, under B. Conner's superintendency. 18 ft x 24 in. Cyls. 6 ft 2 in. D.W. 'improved goods engines'. *Mitchell Library, Glasgow.*

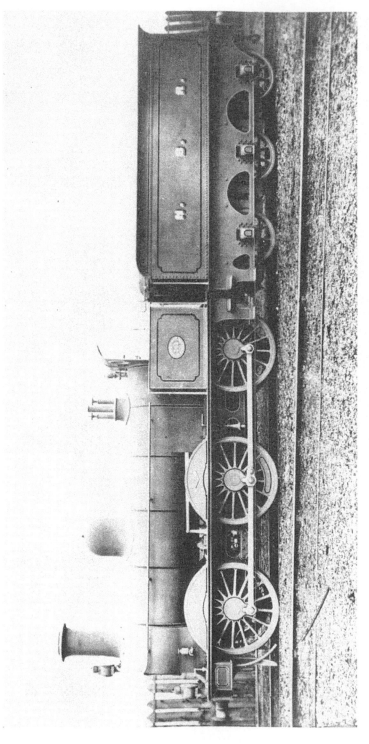

17. North British Railway, o–6–o No. 363, built Dübs & Co., 1869, to designs of Thomas Wheatley. Last engines contract-built during Wheatley régime. *Mitchell Library, Glasgow.*

18. North British Railway, Drummond 0–6–0T No. 96, Cowlairs, 1875.
Railway Magazine.

20. Benjamin Conner, Locomotive Superintendent, Caledonian Railway, 1857–1876.
Adamson, Rothesay,
Copy Montague Smith.

19. William Cowan, Locomotive Superintendent, Great North of Scotland Railway, 1857–1883.
Copy Montague Smith.

with 16 × 21 in. cylinders and domed boilers pressed to 140 lb. per sq. in. In this form they lasted a further 30 years or so, the last to go being No. 216 in 1912.

Another series of Beyer Peacock engines to be given a further lease of life was the '233' class of 2–4–0 first built in 1859. Of these three were rebuilt in 1881 and one in 1882. Again the cylinders were increased in stroke to 22 in., larger boilers fitted, and the pressure raised to 140 lb. per sq. in. Also in 1882 several Beyer Peacock 0–4–2s were rebuilt and given names. These good little engines were again taken in hand by Holmes and given boilers 4 ft. diameter, 9 ft. 7 in. long, with 775 sq. ft. of heating surface, the firebox providing 83·7 sq. ft. The grate area was 15¾ sq. ft. and the working pressure 140 lb. per sq. in. When Holmes's '317' class passenger engines came out in 1903 these 0–4–2s were relegated to the duplicate list and one lasted until 1915.

The departure of Patrick Stirling to Doncaster did not mean the end of Stirling influence in South West Scotland. James Stirling, younger brother of Patrick, who for some years had been works manager, was appointed to the vacant position at Kilmarnock, a post he was to hold for twelve years until he too went south to an English company. James was fifteen years Patrick's junior and was now 31 years of age, two years younger than his brother had been when he took charge of the G. & S.W.R. locomotive, carriage and wagon department, and to date all his career had been spent with this company, first under the elder man's tutelage, then as his subordinate officer. He was therefore well versed in the Stirling tradition and it is not surprising that all ten of his Kilmarnock designs had domeless boilers, safety valves mounted in brass casings on top of the fireboxes, and cabs of the style designed by his brother.

The first locomotives to be built at Kilmarnock to James's designs were fifteen 2–4–0s with 17 × 24 in. cylinders and 6 ft. 7 in. coupled wheels. These first appeared in 1868 and were followed by another group of 2–4–0s, the '71' class, which had 6 ft. 1 in. coupled wheels but were otherwise almost identical with their forerunners, the '8' class, in all other dimensions. The boilers of both classes were comparatively small, having only 779 sq. ft. of tube surface and 82 sq. ft. of firebox with a 15 sq. ft. grate, and pressed to 130 lb. per sq. in. The smaller wheel variety was mainly used on coast and branch line services and one, No. 71, was the first G. & S.W.R. engine to be stationed at Stranraer when the Girvan & Portpatrick Junction line was completed in 1877.

The next three years saw two new classes of 0–4–2, not built this time at Kilmarnock but by contract. First there were twenty by Neilson & Co., in 1870–1, and in 1873 Dübs & Co. built nine. The Neilson engines were a development of Patrick Stirling's engines of 1864–5. Like the '141' class the cylinders were 17 × 24 in. but the coupled wheels were 5 ft. 7 in. diameter and the boilers had slightly more heating surface. The Dübs engines had 18 × 24 in. cylinders and coupled wheels the same size as the Neilsons, and the boilers were similar. The tubes furnished 1,039 sq. ft. and the fireboxes 95 sq. ft. whilst the grates had an area of 15·6 sq. ft. The working pressures were 130 lb. per sq. in. The Neilson engines, '187' class, weighed only 29 tons and their tenders 21 tons 2 cwt. They were intended for fast goods work but did a considerable amount of passenger work as well. Most of this class had lever reverse though some were fitted with the vertical screw reverser developed by James Stirling and fitted to all the Dübs engines which formed the '208' class. In the course of time both classes were fitted with Westinghouse and/or vacuum brakes.

The popularity of the 0–4–2 wheel arrangement was nearly as great as that of the 0–4–0 and the years 1874–8 saw yet another class of 0–4–2s arrive from Glasgow. This was the '221' class, an enlargement of the '208' class. Neilson & Co. built fifty, and Dübs & Co. ten. With 18 × 26 in. cylinders, 5 ft. 7½ in. coupled wheels and larger boilers, they scaled 33 tons 4 cwt. and the tenders 24 tons 14 cwt. The first twenty of the class had vertical screw reverse, the later engines having steam reverse. Manson rebuilt 30 between 1901 and 1903 with domed boilers and his style of cab, and vacuum brake equipment. The result was quite a handsome, businesslike locomotive and as a class they did a lot of very useful work on passenger as well as goods trains. One was involved in the serious buffer-stop collision at St Enoch station, Glasgow, on July 27, 1903, when bringing in a train from Ardrossan.

Stirling produced two classes of 0–4–0, twenty-two tender engines coming from Kilmarnock in 1871–4 and six tank engines in the two years following. The tender engines formed the '65' class and were to be found in various nooks and corners of the system where their flexibility permitted them to penetrate. The cylinders of this class were only 16×22 in. and the wheels 5 ft. diameter. The cylinders of the tank engines were similar and the wheels 4 ft. 7 in. The boilers of both types were no larger than those of the elder Stirling's 0–4–0s of ten years previously.

Just as on the east side of the country Wheatley was preparing for the advent of the Midland traffic through Carlisle, so was James Stirling in the west. The G. & S.W.R. had, until they participated in this new venture, remained a purely bucolic line since it came into being in October 1850. Stirling's one and only 4–4–0 express locomotive proved to be a winner from the start.

The first of these engines came out of Kilmarnock works in 1873 and in the four years following another 21 were built. Like the N.B.R. engines they had inside cylinders and inside frames; Scotland thus led the way with this new type of express locomotive, the first English example not appearing until 1874 when Johnson produced his version on the Great Eastern.

In leading dimensions the G. & S.W.R. engines were rather larger than Wheatley's, the cylinders being 18×26 in. and the coupled wheels 7 ft. 1½ in. The heating surfaces were: tubes 1,028 sq. ft., firebox 84 sq. ft., total 1,112 sq. ft. The grate area was 16 sq. ft. and the working pressure 140 lb. per sq. in. The engines weighed 39 tons and the tenders 24 tons 4 cwt. The leading bogie was typical of the age; like Wheatley's bogies the centre pin was 1 in. in front of the bogie centre-line and formed a fixed pivot without sideplay. The wheelbase was short, 4 ft. 10 in., that of the coupled wheels 8 ft. 3 in. and the total engine wheelbase 20 ft. 3¾ in. Lateral rocking of the engine on the bogie was damped by means of brackets on the main frames and on the bogie frames, one on each side, with rubber cushioning washers between.

When the through Anglo–Scottish services began on May 1, 1876, several of the class, which had been fitted with Westinghouse brake equipment for the purpose, were put on these trains and ran them well. The Ramsbottom safety valves were originally mounted on the firebox, except in the case of the prototype engine, No. 6, which had balanced valves in a brass casing. No. 6 also had the Stirling vertical screw reverser when new. Both fittings were later changed to conform with the rest of the class which had the steam reverser.

The Stirling method of entraining the steam from the boiler to the cylinders was by means of an internal pipe running the whole length of the boiler, perforated at the top, and terminating in a regulator box, or header, in the smokebox. The header was of cast-iron and contained two horizontal slide valves operated by a long rod from the footplate, with the usual stuffing box on the boiler back-plate. In the smokebox a pipe led from the header to the steam-chest. The Stirling claim for this method was that no large hole in the boiler barrel for the dome was required, and the barrel was thus stronger, and water could be readily separated from steam by means of the perforated pipe without the need for a dome. Further, manufacturing costs were reduced.

After some 23 years these racy looking engines were to be found on local passenger trains and piloting jobs and were on the duplicate list; nevertheless, at the turn of the century, Manson rebuilt sixteen of them, giving two of them domed boilers. The tube surface was reduced to 868 sq. ft. and the firebox increased to 95 sq. ft. The pressure was raised to 150 lb. per sq. in. The weights of engine and tender were increased by one ton and half a ton respectively. In their new guise they lasted well into L.M.S. days and did some good work.

Two other classes were designed by Stirling, and work began on them before he left the South Western, though it was for Smellie to follow their construction through.

First there were the twelve engines of the '13' class 0–6–0 goods engines, an adaptation of the '221' class 0–4–2. Dimensionally similar, save for 5 ft. 1 in. coupled wheels instead of 5 ft. 7½ in., the 0–6–0 was a stronger engine and permitted a 20 per cent greater load than the '221' class. When built these engines carried the hall-mark of Stirling's successor—the safety valves were moved to a position in the middle of the boiler.

The second class commenced by Stirling and finished by Smellie was, strangely, a tank engine built specially for the Greenock road which was steeply graded. For some reason never satisfactorily explained, tank engines were anathema on the G. & S.W.R. Certainly there had to be some for shunting duties but even so, much of this work was done by tender engines. Even the City of Glasgow Union line was operated by tender engines, jointly with the N.B.R. since its opening in December 1870. The idea of running the whole distance to Greenock over the heights of the Renfrewshire hills must have shocked some people. However James Stirling tried it, though he got little more success than might have been expected on this railway with its aversions to tank locomotives.

The 0–4–4T he produced at Kilmarnock in 1879 was a neat little design having 18 × 26 in. cylinders, 5 ft. 7 in. coupled wheels, 932 sq. ft. of tube surface, 93 sq. ft. of firebox and 15 sq. ft. of grate. Pressed at 130 lb. per sq. in and weighing 45 tons 16 cwt. they could have been a very useful class but for the inherent prejudice of the enginemen. The side tanks carried 1,000 gallons of water and the rear bunkers 1½ tons of coal. There were complaints that the engines rolled too much when running downhill and that they did not carry enough water. To some extent this can be understood. The hard work climbing the steep bank out of Greenock up to Port Glasgow would account for nearly one third of the water content. This together with the coal consumed would so reduce the weight on the springs that rolling can be readily imagined. Against these criticisms must be placed the enginemen's natural, or unnatural, prejudices. Unpopularity on the Greenock road caused them to be put on the suburban services to Barrhead, Potterhill and Johnstone. When Whitelegg embarked on his post-war make-do-and-mend in 1920 he gave them larger bunkers and sent them to places like Ayr, Beith, Girvan etc. Such was the '1' class of 0–4–4Ts and the four little engines lasted a very long time, certainly well into the L.M.S. era.

On April 2, 1878, James Stirling's resignation was accepted by the board as he had been successful in obtaining a similar post on the South Eastern Railway at Ashford in Kent, and on April 30th his successor, Hugh Smellie, was appointed.

David Jones, born in Manchester in 1834, arrived at Inverness a few days after the opening of the Inverness & Nairn Railway on November 5, 1855. He was soon engaged by that company as an engine-driver, and his abilities were quickly appreciated by the board so that in 1858 he was appointed assistant locomotive, carriage & wagon superintendent under William Barclay.

Jones had served an apprenticeship under John Ramsbottom at Longsight on the L. & N.W.R., and was thus no stranger to the 'Old

Crewe' types of locomotives he found at Inverness. Now, at the age of 36, he was in charge of the department he had helped to build up from the meagre beginnings of 14 years before.

During this time he had seen the growth of the small companies which had amalgamated to form the Highland Railway and further development was projected. The Caithness Railway to Wick and Thurso had been authorized and constructional work begun. The extension of the Ross-shire Railway to Bonar Bridge was opened on October 1, 1863, the Sutherland Railway thence to Golspie on April 18, 1868, and work was in hand in the making of the Dingwall & Skye Railway which was opened on August 10, 1870, eight weeks after Jones had taken office.

A condition of his appointment was that any additional duties devolving on him as a result of the opening of the Skye line and the Sutherland Extension, would be covered by the salary of £500.

The state of the engine power of the Highland in 1869 was, on the whole, fairly good. There were just about enough engines, mostly 2-2-2 and 2-4-0 types, though some were now beginning to show an incapacity to cope with the growing needs. Jones was fully aware of these shortcomings, and he at once set about programming a series of improvements. In early 1870 however, there had not been a full recovery from the financial depression of the middle sixties and money was still tight. New work would therefore have to be shelved for a while, but some improvement was necessary. It was quite evident that the 2-4-0 engines were superior to the singles, at least on the main line with its two summits at Dava and Druimauchdair, so a start was made by rebuilding some of the 2-2-2s as 2-4-0s, just as Stroudley had done with *Raigmore*. Between 1871 and 1879 nine were so treated but, whereas Stroudley only increased the cylinder from 15×20 in. to 15½×21 in. Jones went one better and gave them 18×24 in. cylinders and improved the boilers by giving them 1,093 sq. ft. of tube surface, 93 sq. ft. of firebox, removing the mid-feathers, making a total of 1,186 sq. ft.; the working pressure was raised to 160 lb. per sq. in. The locomotives so treated were all Hawthorns and Neilsons built in 1863-4 and the cost of the alterations was no more than £300 per engine. The increased cylinder sizes were evidently a success, for some of the earlier 2-4-0s were likewise dealt with between 1872 and 1876. In 1871 one of the Hawthorn singles of 1862 was rebuilt as a tank engine but changed dimensionally, and put to work on the Aberfeldy branch, being named *Breadalbane* in accordance with Highland custom. Names were bandied about with the same *sang-froid* as on the North British, though quite often spelling errors would appear. If engines were sent to shops in the south, as they were after 1923, there could be some excuse, but there was no excuse for such mistakes made by Highlandmen in Highland works in Highland territory.

The same year, 1871, saw No. 2, *Aldourie*, one of the two original I. & N.R. singles withdrawn and scrapped. Such parts as could be used were incorporated in a new locomotive, similarly named, which had new frames supplied by Hawthorn & Co. and a new boiler from Dübs & Co.

Very soon after the opening of the Dingwall & Skye Railway it became apparent that the 2–4–0 wheel arrangement was too rigid for the sharp curves on this line. Accordingly Jones took No. 10, one of the 1858 Hawthorns, into Lochgorm works, removed the leading axle, modified the fore end framing, and fitted an Adams bogie. At the same time the cylinders were enlarged to 17×24 in. No alteration was made to the boiler. The engine weight was increased from 28·47 tons to 32·5 tons and an immediate success was obtained with the more flexible wheelbase. The rebuild of No. 10, and the manner in which Cowan's 4–4–0s on the G.N.S.R. were behaving contributed in no small measure to Jones's decision to adopt this wheel arrangement for his next new locomotives, which were now badly needed despite the improvements already effected in the older engines.

An order was placed with Dübs & Co. in Glasgow for ten 4–4–0 engines and tenders, the first three of which were delivered in June 1874, followed by four in July and three in August. The numbers were 60–9 and their dimensions made them the most powerful engines on either of the northern companies. Jones followed tradition by using the Allan fore end; the cylinders were 18×24 in., the coupled wheels 6 ft. 3 in. and the pressure 140 lb. per sq. in. Thus the tractive effort was 12,337 lb. at 85 per cent B.P. The boilers were also improved having $223 \times 1\frac{3}{4}$ in. brass tubes giving 1,132 sq. ft. heating surface. The firebox, which had a horizontal grate 16·25 sq. ft. in area, had a surface of 96 sq. ft. making the total

1,228 sq. ft. There was therefore a good capacity for generating steam. Again Adams bogies were fitted. In these the centre bearing was taken on a rubber pad inserted between the horizontal slide and the frame stretcher on the underside of the smokebox. Side-play was controlled by rubber springs. The valve motion was Allan's straight link and the slide valves were of brass. To reduce the danger of fire in the thickly wooded country through which the line passed a wire spark arrester was fitted between the blast pipe orifice and the base of the chimney, which was of Jones's louvred pattern. Careful thought had been given to the springing. The main lines of the Highland passed over a variety of geological formations. In one place the road would be laid on native rock, in another, it might be on near bog. To allow the engines to accommodate themselves to such variations, especially when running downhill, the coupled wheel springs were fitted with compensating beams so that the weight was taken on the bogie at the front, and on the coupled wheel axleboxes on each side at the rear, a good example of three-point suspension. The weight of these engines when built was 41 tons with 26½ tons available for adhesion; a small amount of this was obtained by means of a small cast-iron water tank under the footplate forming a well between the tender and the injectors. After the 1875 Newark Brake Trials the Highland company adopted the vacuum brake, and when this was fitted to these locomotives it increased the weight by one ton. The tender carried 4 tons of coal, 1,800 gallons of water and weighed 30 tons. Additional engines of the class were built in the company's own works at Inverness, the first being No. 4 in 1876. In 1883–8 another six were added but these carried a slightly modified boiler pressed to 150 lb. per sq. in. In these the heating surfaces were reduced, the tubes by 74 sq. ft. and the fireboxes by 3 sq. ft. Of the older engines two were given similar boilers, No. 4 in 1895 and No. 67 in 1897.

In 1875 another of the Hawthorn singles was rebuilt to the same dimensions as No. 10, and this one, No. 7, also went to work on the Skye line.

At this period in the history of the Highland Railway there were only five tank engines in the stock: No. 17, the 0–4–0T Stroudley had built out of the Hawthorn 0–4–0T of 1863, Jones's conversion of that firm's single No. 12, in 1871, and the three 'R' class o–6–oTs begun by Stroudley and finished by Jones, Nos. 56, 57, and 16.

To meet the requirements of branch lines Jones built three 2–4–oTs of modest size at Lochgorm works in 1878–9. They were originally intended for station shunting and had 16×24 in. cylinders and 4 ft. 9 in. coupled wheels and the leading wheels were 3 ft. 9½ in. The heating surfaces were: tubes 820 sq. ft., firebox 93 sq. ft., total 913 sq. ft. and the grate area was 16·2 sq. ft. The working pressure was 140 lb. per sq. in. and the weight 36 tons 11 cwt.

Trouble arose in service from the rigidity of the leading axle and in 1881 two, and in 1882 the third, were rebuilt as 4–4–oTs by the simple expedient of replacing the leading axle by a bogie as had been done with the two tender engines. In the case of the tank engines the bogie wheels were only 2 ft. 7½ in. diameter in a 5 ft. 6 in. wheelbase. The weight was increased to 39½ tons. These were good looking and hard working engines that had a long life, the first to be broken up going in 1928 and the last five years later.

The popular 4–4–os gained a further hold by the advent of a new class in 1882. These were based on the two conversions and the '60' class, but were given 5 ft. 3 in. coupled wheels specially for the Skye line. The boilers were 1 in. shorter than those of the '60s', otherwise the dimensions were similar. Weighing 43 tons, with 28 tons on the coupled wheels, they had tenders similar to the '60s', and the working pressure was 150 lb. per sq. in. Lochgorm works built nine altogether, the first, No. 70, appearing in 1882. Not until 1892 was another built, and this was followed by two in 1893 and a fifth in 1895. No more were built during Jones's superintendency but under Drummond four came out between 1897 and 1901.

On the Great North of Scotland Railway William Cowan was still in command at Kittybrewster but had made no additions to the locomotive stock since the 'K' class of 1866. However in 1876 the need for new engines resulted in six 4–4–os being obtained from Neilson & Co. The '49' or 'L' class handled nearly every kind of traffic on the G.N.S. and appears to have done so very efficiently. Slightly more powerful than their predecessors, they had 17×24 in. cylinders, 5 ft. 7 in. coupled wheels and weighed 39 tons 14 cwt. The heating surfaces were: tubes 1,023 sq. ft., firebox 84 sq. ft., total 1,107 sq. ft., and the grate area and working pressure were 14 sq. ft. and 140 lb. per sq. in. respectively. Pickersgill rebuilt these engines in 1897 when the pressure was raised by 10 lb. per sq. in. and the weight increased to 41 tons 4 cwt. Like all G.N.S. locomotives they had the Clark smoke consuming apparatus and a brick arch was also fitted in the firebox. The bogies were an improvement on those of the 1866 engines in that they had the more orthodox centre pins and flat bearing plates. The characteristic Clark design was apparent in all these engines and was carried on to Cowan's last two products, the 'M' class, which was brought out two years later in 1878, and the 'C' class of 1878-9. In both instances Neilson & Co. were the makers, and there was a considerable degree of similarity between them. Both had cylinders $17\frac{1}{2} \times 26$ in. and whereas the 'M' class, starting with No. 57, had 5 ft. 7 in. coupled wheels, those of the 'C' class, Nos. 1, 2, and 3, were 6 ft. 1 in. The boilers of all three classes were alike, but there was a slight difference in the weights of engines and tenders, though none could be called heavy.

Although the 'M' class was projected when the board was calling for larger and more powerful engines, it would seem that they were not prepared to face the increased costs of operation, despite the reduction in coal consumption claimed for Clark's device already described. This attitude on the part of the Directors was by no means uncommon on the Great North, and as in the cases of Clark and Ruthven Cowan became the target for the board's testiness. The outcome of the ill feeling bred by this attitude was that in 1880 Cowan resigned. The same year saw the appointment of William Moffatt of the North Eastern Railway as general manager, A. G. Reid as superintendent of the line, while William Ferguson of Kinmundy became Chairman of the company. These changes resulted in a considerable improvement of the company's affairs in general, but particularly in the timing and time-keeping of trains, and traffic operation as a whole.

On the death of Benjamin Conner in February 1876, the man who had been his assistant and had deputized for him during his last illness, George Brittain, was appointed in his place.

At this time St Rollox was preparing designs for a new 2-4-0 and an order for five had been placed with Neilson & Co. When the new engines arrived they were 4–4–0s! During his service life Conner had been a devotee and advocate of Allan's designs, but the new engines had nothing of the Allan about them. For one thing, Conner would not countenance the use of a leading bogie, preferring the 2–4–0 arrangement of which he had constructed so many. How the change was brought about has not been recorded, but Messrs Neilson's Order Book bears the following notes, 'Bogies to be substituted for leading wheels as shown on tracing 9983 submitted by us on 24/1/76 and approved by their letter of 24/1/76.' And again, a marginal note, 'Mr Conner dead'. From that it would seem that Brittain, acting for his chief, agreed the alteration while Conner was on his death-bed, and that the second note was added later. It is perhaps as well that Conner did not live to see these engines in service; they were failures. The cylinders were 18 × 24 in. and the coupled wheels 7 ft. 2 in. The bogie wheels were 3 ft. 4½ in., the axle centres being 6 ft. apart. Once again the bogie centre pin was 1 in. in front of the bogie centre line. The heating surfaces were: tubes 905·26 sq. ft., firebox 82 sq. ft., total 987·26 sq. ft. and the grate area was 14·6 sq. ft. The working pressure was a modest 140 lb. per sq. in. The weights were: bogie 12 tons 14 cwt. 2 qr., driving axle 15 tons 7 cwt. 2 qr., and trailing axle 13 tons 5 cwt. The six-wheeled tender carried 4 tons of coal and 1,880 gallons of water and weighed 29 tons 7 cwt. 1 qr. The total weight of the engine and tender was thus 70 tons 14 cwt. 1 qr. The Caledonian practice of providing a hand-rail on the side of the tender and a footstep from front to rear of the tender at axlebox hornstay level, for the use of shunting staff, was followed but after the retirement of Brittain did not appear again.

These engines went straight on to the Anglo–Scottish trains from Glasgow and Edinburgh, but put up such poor performances that they were soon transferred to the Glasgow and Dundee local services. A slight improvement in steaming propensities was obtained when Drummond rebuilt them with his own boilers, and when the Lanarkshire & Ayrshire Railway was opened in 1890 some were put on the Ardrossan trains.

Two features of these engines are worth noting. First, Conner was evidently so imbued with the perfections of the Allan designs, and so set in his ways, that he could see no good coming from a departure from these principles, so that it was left to his successor to lift the Caledonian out of the slough into which it was in danger of falling, and once for all eschew the Allan tradition. Second, by the use of the Gooch valve gear, the boiler centre-line was kept down to 6 ft. 9 in. above rail level even though the driving wheels were over 7 ft. diameter.

An interesting little engine built by Neilsons in 1878 was a crane engine for use in St Rollox works yard. It was a strange looking contraption which had the crane jib, working at a fixed radius of 8 ft. 6 in. mounted on the smokebox, the chimney, extended in height to clear the jib, forming the king-post of the crane. The lifting capacity was 2 tons. The side tanks, for 200 gallons, were set well back, and the coal bunker carried 5 cwt. so that there was a fair counterbalance for the weight forward. Large dumb buffers were provided, as well as a cab consisting of a front weatherboard and a roof supported by pillars. Other dimensions were: inside cylinders 11×20 in., 3 ft. 3 in. wheels, 439 sq. ft. of heating surface and 7 sq. ft. of grate. The weight was 17 tons 17 cwt.

In 1878 and again in 1881 Dübs & Co. built some 0–4–2 engines for goods work. Somewhat similar to the 1870–1 batch in many respects they had a greater heating surface by 320 sq. ft., i.e. 1,118 sq. ft. The grate area remained the same 13·3 sq. ft. and the working pressure was 140 lb. per sq. in. After the failure of the 2–4–2T engines on the Callander & Oban line this '670' class took over.

Branch line workings were catered for by Brittain's 2–4–0 tanks of which Dübs & Co. built twelve in 1879. Weighing 41 tons 10 cwt., they had outside cylinders 17×22 in., 5 ft. 3 in. coupled wheels, 1,070 sq. ft. heating surface, 13 sq. ft. grate area and 140 lb. per sq. in. pressure.

The next engine Brittain designed was a 2–4–2T specially for the Callander & Oban Railway which the Caledonian was working, though to all intents and purposes it was a separate entity. Neilson & Co. built fifteen of these locomotives in 1881. In his *Callander & Oban Railway* John Thomas has told very clearly the story of their advent and has exploded the theory that several writers have expressed, that these engines were the outcome of trials with one of Webb's 2–4–2T from the L. & N.W.R. That Brittain was influenced by Webb's work seems evident from his use of a rather similar axlebox of the radial type for the leading and trailing axles, but at the date Brittain's 2-4-2T engines were on the drawing-board, only the 2-4-0T engines with radial axleboxes had been turned off Crewe shops. Webb's 2-4-2T appeared in 1880 when, as Mr Thomas has related, Brittain's engines were already under construction. One of Webb's 2-4-2Ts was certainly tried on the Oban road late in 1880, but this was after the debacle caused by the repeated derailments of the St Rollox design, and it may well be that it was in an attempt to determine the causes of these mishaps that a comparison with the Crewe product was arranged. That these tests did not result favourably for the Webb design seems to be borne out by the fact that no more were built by the Caledonian, or by any other Scottish railway.

The dimensions of the 2-4-2T were as follows: cylinders $17\frac{1}{2} \times 22$ in., radial wheels 3 ft. 8 in., coupled wheels 5 ft. 8 in., $210 \times 1\frac{3}{4}$ in. tubes 1,010 sq. ft., firebox 81·5 sq. ft. totalling 1,091·5 sq. ft., grate area 13·4 sq. ft. The pressure was 130 lb. per sq. in. and the tank and bunker held 1,200 gallons and $2\frac{1}{2}$–3 tons respectively. The total weight was 51 tons 12 cwt. of which 29 tons 17 cwt. rested on the coupled axles. Withdrawal of the class finally was begun in 1899 as a result of their unreliability to remain 'on the road', even after the radial axleboxes had been blocked when the engines were relegated to suburban work in the Glasgow area. Most of them had been scrapped by 1912, though one did last as late as 1922. The numbers borne by these engines when new were 152–66.

There was certainly a need for better engines, and until such time as the drawing office and the builders could produce something new, the '670' class of 0-4-2s was drafted in to work the traffic on the Oban line. Then, in 1882, from Dübs & Co's works came ten excellent little 4-4-0s. In appearance they were very similar to the 'Skye Bogies' of the Highland Railway which were under construction at about the same time. However, whilst the latter faithfully adhered to traditional Allan fore-end, Brittain simplified the design. The cylinders were the same bore and stroke as the Highlanders, 18×24 in., but the coupled wheels were 1 in. less, i.e. 5 ft. 2 in. diameter. There was considerable difference in the heating surfaces, the Caledonian engine having only 1,053·22 sq. ft. of tube surface, nearly 70 sq. ft. less than the northern engine. The fireboxes were more nearly alike, the Caley engine having 93·2 sq. ft. Thus the totals were Caledonian, 1,146·42 sq. ft., Highland 1,228 sq. ft. and the grates were 14·4 sq. ft. and 16·25 sq. ft. respectively. The Caledonian engine carried a working pressure of 130 lb. per sq. in., 10 lb. per sq. in. less than the Highland and the tractive efforts were distinctly in favour of the northern engine at 14,687 lb. compared with 13,858 lb. The weight of the 'Oban Bogie' as the class became known was 41 tons 11 cwt. 3 qr., of which 14 tons 4 cwt. was on the bogie. Four-wheeled tenders of 4 tons and 1,550 gallons capacity, weighing 24 tons 1 qr. were fitted with handrails and footsteps. Between 1898 and 1901 McIntosh rebuilt the whole class, which were numbered 179–188, with boilers of 1,089·89 sq. ft. surface, made up of 975 sq. ft. tubes, 110 sq. ft. firebox and a 17 sq. ft. grate. The working pressure of McIntosh's boilers was 150 lb. per sq. in.

In service the class proved useful and capable engines, handling the traffic, which at holiday seasons could be very heavy, on this steeply graded line in excellent manner until towards the close of the nineteenth century train loads proved too much for them without the aid of a pilot engine.

The three little 2–2–2T engines built at Greenock in 1851, contemporary with the 'England' types of very small engines of the early fifties, were broken up in 1879–80 and the serviceable parts used to make a rebuilt No. 1, and a 2–2–2T, specially to haul the directors' saloon, the Caledonian counterpart of the N.B.R. 'Cab'. This miniature locomotive had 9×15 in. cylinders, 5 ft. 1 in. drivers and a wheelbase of 13 ft. In working order the original engine had weighed 26 tons 14 cwt. The springing was unusual, plate springs being applied to the leading and trailing wheels and a transverse plate spring to the drivers. It was given Gooch's valve gear, outside, and like most tank engines of the fifties had brake blocks to the trailing wheels only.

In 1882 Brittain's health was failing and for this reason he resigned in April of that year. At the same time as a successor was being considered, opportunity was taken to carry out some reorganization of the department.

Minute No. 858 of the Directors' meeting of June 20, 1882, records the appointment of Dugald Drummond as locomotive superintendent at a salary of £1700 per annum. When appointed to the North British Railway in 1875 he was paid £700, until

March 1876 when his salary was raised to £850 for one year and then to £1000. His successor Matthew Holmes, was only to be paid £750; on the Highland Jones got £500, and on the Caley Brittain was paid £700 when he took over in 1876; Drummond's salary was raised to £2000 as from August 1, 1884!

Some of the design features of the period are worth brief consideration. Main frames were generally of best Yorkshire iron, and double frames were gradually giving place to single, inside frames which were growing in popularity. At Cowlairs Drummond was making his frames in one piece from the back buffer beam to the front end of the cylinders where they were welded. The front portion, including the front buffer beam, was forged in one piece and welded to the frames. Both Jones on the Highland, and Drummond on the N.B.R. were using the horseshoe pattern of hornblock, the former of cast steel and the latter of good tough cast iron.

Most boiler barrels comprised three rings but here again Drummond was leading, for the Cowlairs practice in his time was to use two rings only, butt jointed with butt straps inside and outside, double riveted. Lap joints were usually single riveted. Again the best quality Yorkshire, usually specified as Low Moor, iron plates were used. It is noticeable that in Drummond's 4–4–0T of 1879 the firebox foundation ring is only two inches wide, but the width of the water spaces increased towards the upper parts. Some boilers, like the 'H' class 4–4–0 of 1862 on the G.N.S.R., had the longitudinal barrel seams along the bottom. No. 31 burst her boiler at Nethy Bridge in 1878 due to corrosion along the seams. As a result the working pressure of the whole class was dropped to 100 lb. per sq. in. until new boilers had been constructed.

On the G.N.S.R. Cowan was also responsible for the fitting of an early form of drop grate of his own design. The firebars, which were able to move in a vertical plane, about their back ends as fulcrum points, could be raised or lowered at their front ends by means of a screw gear in the cab. When lowered, and with the front damper open, the clinkers could be pushed out through the ashpan.

The staying of firebox crowns varied. Jones used three girder stays on each side of the crown plate with two rows of direct stays in the centre, while Drummond had two rows of sling stays, hung from a 'T' angle riveted to the inside of the wrapper plate, and as many rows of direct stays as were necessary for the length of the box.

On the G. & S.W.R. Patrick Stirling's locomotives received water into the boiler through clack boxes immediately above the foundation ring. This arrangement did not prove wholly satisfactory, probably on account of resultant plate wastage in the vicinity, and he reverted to putting the clack boxes on the boiler barrel about midway between the tubeplates and on the centreline. Drummond put his clack boxes further forward, about 1 ft. behind the smokebox tubeplate, and Conner put them midway between the two.

Safety valves have been mentioned several times. The Ramsbottom pattern was becoming more popular though the Salter balance type was still common with the northern companies. Jones was, however, replacing these and the Naylor valves with his own lock-up valves and he also tried out the Adams 'pop' valves, which formed the basis of the 'pop' valves of later years. Naylor valves were found to be unreliable because of their prevalence for sticking and, when this occurred, their inaccessibility. Cowan continued the Clark practice of putting the valves on the dome and both on top of the firebox. James Stirling put his Ramsbottom valves here until they were moved by Smellie to mid-barrel, making a more pleasing appearance than was afforded by the severely functional, unadorned barrel of the Stirlings.

At the other extreme, Drummond's practice of putting his Ramsbottom valves on top of the dome, mid-way along the barrel, albeit with a short easing lever only, tended to give his engines a cocky, slightly overdressed appearance. However, such was their excellence in service that such a small conceit could without doubt be condoned.

The round section connecting rods of early engines had almost completely disappeared, at least on new work they were no longer used and they only existed on some of the older engines still in service. Big ends were mainly of the gib-and-cotter type, though Drummond was using the marine type with considerable success. On four-coupled types coupling rods were usually bushed but on six-coupled engines the usual arrangement was split brasses and cotters.

Valve gears were of various kinds. The Allan straight-link was a great favourite on the Highland and there were a few instances of it on other Scottish lines. The Gooch was to be found on a few engines but the most popular was the Stephenson link motion. It was used by most companies except the Highland.

Cabs were being provided on all engines after the start of the seventies. The Barclay/Stroudley cab of the early Highland engines was developed by Jones into a really good protection for the enginemen, well deserving the term 'house' which was given to it by the men. With its rounded corners the Jones version had a very pleasing appearance. The Stirling cab did not afford so much protection for the men but was better than nothing. A form of this cab was used by Drummond on his rebuilds of the E. & G. engines and even later it was used by Holmes. Wheatley's poor little bent-over weather board did not last long, and Conner's Caledonian cabs were not much better. By far the best cab was that designed by Drummond. It was based on the Stroudley pattern but carried the unmistakable stamp of the Drummond school of thought and set a pattern for many years to come. There was an extraordinary reluctance on the part of the men to take to cabs, brought up as they were to brave the worst elements stoically, and on management side there was a similar reluctance on the part of many engineers to provide such protection on the grounds that it would tend to diminish the keenness with which the footplate staff carried out their duties. Commonsense prevailed and before the 80s were out it was the exception to find an engine without reasonable protection for the men.

The Stirling steam reverser was a standard fitting on the G. & S.W.R. It consisted of two cylinders placed in tandem with a common piston rod. This rod was connected, usually at the centre, by linkage to the reversing gear. One cylinder contained steam, the other oil. To reverse the engine a steam valve was opened, allowing steam to enter one cylinder and move the piston and oil which flowed through a valve to the other side of the piston. This movement caused the reversing arm to alter the position of the quadrant link thus reversing the engine. The oil kept the piston in any desired position. As soon as oil could not pass from one side of the piston face to the other the gear was firmly locked. One drawback was that when an engine was stabled in the shed it was necessary to see that it was left in the gear required for the next move, or mid-gear, otherwise it would be necessary to await sufficient steam being raised to enable the gear to be operated.

On the G.N.S.R. and N.B.R., including the E. & G.R. engines, the smokebox doors were hinged at the left-hand side and secured by means of a dart and crossbar, in the manner that became almost universal. The early Highland and Caledonian, with their constituents, had, like so many of the Allan, or 'Old Crewe' type engines, the top-hinged door, but with the coming of Jones and Conner the side-hinged door became more generally used. The G. & S.W.R. had almost invariably used this type of door. From the point of view of access to the tubes for cleaning, or to steam- or blast-pipes for repairs, the side-hinged door was vastly preferable.

Chimneys showed some variety. On the N.B.R. the parallel chimney with flared top gave place to Wheatley's stove pipe which was very similar to the one Conner was using at the establishment across the road. Wheatley preferred the plain unadorned style, and in all his new engines he used it, giving it a slight outward taper towards the top, as did Conner. On the other hand the G.N.S.R. was fitting chimneys with the taper the other way, i.e. they were slightly wider at the base than the top. The modern chimney with its throat near the base is the development of the former and the result of an appreciation that the exhaust steam is still expanding and filling the bore of the chimney thus creating the necessary piston action in the chimney which ensures a good draught on the fire.

The Highland, in an effort to minimize the risk of fires in forested areas, and there were many in its territory, adopted an entirely different arrangement. The chimney devised by Jones consisted of an inner chimney proper, entraining the exhaust steam and smoke from the smokebox, and an outer chimney which had a number of louvres in the front semi-circumference. The object of this arrangement was said to be two-fold. First, to create an upward draught in the annulus between the two chimneys so that the exhaust, and any sparks lifted with it, would be thrown to such a height that they would be innocuous when they came down, and second, to keep the cab spectacle glasses clear of down-driven smoke when running with little steam on and a soft exhaust, or with the regulator closed.

When Dugald Drummond arrived on the scene at Cowlairs he

brought out his own distinctive chimney, described by one writer as a 'muzzle loader'. This also had an outward taper towards the top, about 1 in 30, and the top was fitted with a cast-iron cap.

Tenders were usually either four-wheeled or six-wheeled. More usually those engines required to negotiate tight curves in colliery areas, dock-yards, etc., would have the former; main line loco- motives were generally provided with the six-wheeled variety. Both types of tender were constructed with double frames and with single frames. All four kinds were to be found on the E. & G. and N.B. Railways. On the Caledonian, Conner used single and double frames whilst Brittain preferred inside, and therefore single, frames. The G.N.S.R. was employing outside-framed tenders with four and six wheels until the late seventies when the inside frame arrived on the 'M' class from Neilsons.

Where outside or double frames were employed springing was a matter of some consequence. Most tenders of this type had the springs between the frames, and thus most inaccessible for erection in the works, and particularly for changing broken or weak springs by shed fitters when in service. The change to inside frames spelt considerable relief in this respect. The first development was to have the springs above the narrow frame platform with the hangers and pillars passing through this to the pins and axleboxes below. Later the practice of mounting the springs direct on to the axlebox tops, and therefore beneath the frame platform, became commonplace, though Drummond on his 'Abbotsford' 4-4-0 on the N.B.R. followed the Stroudley practice of slinging the springs underneath the axleboxes. This prevented the use of these tenders on the Forth Bridge when it was opened in 1890, because they interfered with the troughing in the road-bed.

On the G. & S.W.R. tenders were very much the same as has been described for other companies. In one respect some of them differed. These were the Stirling tenders which had a horseshoe- shaped tank, with the coal space between the arms of the horseshoe instead of the customary practice of piling it on top of the sloping tank top. These Stirling tenders were coupled to the engines by a screw coupling with spring-loaded side buffers as well. When the springs were worn and weak a jigger-screen movement could be initiated, by no means improving the riding of the engines besides covering the footplate with coal.

Chapter VI

THE TRENDS DEVELOP

Hugh Smellie returned to Kilmarnock at a very interesting period. His predecessor's high-stepping express engines were swinging their ways down 'f'ae the Shilford tae the Gorbals' and around the Ardoch curves with the Pullman trains, and the continuous brake experiments were well in hand. The 'Black Goods' were earning freight revenue on the long-haul goods trains and even the unliked 0–4–4T on the Greenock road were playing their part despite their tendency to roll and their alleged thirst for water.

Smellie's first engine for the South Western was a 2–4–0. Some surprise has occasionally been expressed at his choice of this wheel arrangement in the year 1879. The opinion has been put forward several times that he was distrustful of the new device of a leading bogie. But there was nothing new in the use of a leading bogie for express locomotives. It had been introduced into Scotland by Cowan on the G.N.S. in 1862. Its only claim to newness was in conjunction with inside frames and inside cylinders, and of this type there were now several in the country. The N.B.R. had at least three types, the G.N.S. and Highland Railways had some but the Caledonian, like the G. & S.W. had so far only produced one class each. These engines gave no cause for anxiety regarding their riding propensities and the companies were thoroughly satisfied with the performance of their engines.

There must have been some other factor which caused the new chief to select the 2–4–0 style for his first design, and it seems to have been occasioned by an operational difficulty. Some of the turntables, e.g. that at Stranraer, were too short to turn a 4–4–0 and the new engines would be needed to work between Glasgow and Stranraer on the services made possible by the opening of the Girvan & Portpatrick Junction Railway in 1877, and on the Carlisle–Stranraer services over the Port Road, as well as on the main line.

Whatever the reason, the new engines, when they came out from Kilmarnock works in 1879, were winners from the start. The cylinders were 18 × 26 in. and the coupled wheels 6 ft. 9½ in., which became the standard diameter for the coupled wheels of G. & s.w. express engines for the next 24 years. The leading wheels were 4 ft. 4½ in. diameter; their axleboxes were fitted with inclined planes on the Cartazzi principle and were given ½ in. side-play. These boxes seem to have given a lot of trouble due to overheating of the bearings. The leading hornblocks of the coupled wheel axleboxes were provided with adjusting wedges to take up the slack when wear occurred and prevent undue knock. The wheelbase from leaders to drivers was 8 ft. and thence to the trailers 8 ft. 3 in.

The boiler, which was 4 ft. 2 in. diameter, contained 240 tubes 1⅝ in. diameter giving 1,105 sq. ft. of heating surface, while that of the firebox was 101 sq. ft. making a total of 1,206 sq. ft. with a grate area of 16 sq. ft. A working pressure of 140 lb. per sq. in. allowed a tractive effort of 12,300 lb. at 85 per cent B.P.

In his later designs James Stirling had used valve rods of rectangular section, suspended by swing links but Smellie changed this practice and substituted valve spindles of circular section which worked in bushed guides mounted on the motion plate. He did this not only on his new engines, but on the 0–4–4Ts which were under construction on his arrival. The weights of the 2–4–0s were 11 tons 12 cwt. on the leaders, 14 tons on the drivers and 12 tons 18 cwt. on the trailers, thus the total weight was 38 tons 10 cwt. The first four came out in 1879 followed by eight more in 1880. They were known throughout the line as the 'Twelve Apostles'. Only two remained to be taken into L.M.S. stock in 1923.

At this time the South Western had only some 44 engines of the 0–6–0 type. Traffic was increasing and there was, moreover, a new traffic developing due to interchange with the Midland which had at last arrived at Carlisle. Smellie's next task was to design an 0–6–0, and in 1881 the first of eight new goods engines emerged from Kilmarnock works. They were followed at intervals until the class numbered 64 engines. Kilmarnock built eight in 1885, eight in 1888, four in 1889, six in 1890, and ten in 1891, whilst Dübs & Co. built ten in 1889 and Neilson & Co. ten in 1883.

The cylinders were 18×26 in. and the wheels 5 ft. 1¼ in. Boiler dimensions were rather less than Stirling's '13' class of 1877, the tube surface being only 956 sq. ft. provided by 195×1¾ in. tubes, the firebox gave 105·2 sq. ft. and the grate area was 16 sq. ft. and the boiler pressure 140 lb. per sq. in. The leading wheels carried 12 tons 13 cwt., the drivers 13 tons 5 cwt., and the trailers 11 tons 10 cwt., making a total of 37 tons 8 cwt. The first engine out of the works was No. 22 which gave its name to the class though they were more generally known as the 'Steam-brakers' since they were the first G. & S.W.R. engines to be built and fitted apart from the early experiment by Robertson in the early G.P.K. & A. days.

In 1903 twelve of the class were reboilered by Manson who gave them domes, then in 1911–12 he performed a major operation on them by raising the pressure to 150 lb. per sq. in. and the boiler centre line to 7 ft. 3 in. above rail level. The heating surfaces and grate area were likewise altered slightly and the weight increased by 1·1 tons. The tenders weighed 27 tons and carried 4 tons of coal and 2,200 gallons of water. An example of the work of this class is given by D. L. Smith in his *Tales of the G. & S.W.R.* He cites one regularly employed on a mineral train from Muirkirk to Elderslie all the way round by Auchinleck, Mauchline, Hurlford and Dalry, with a load of 33 equal to nearly 400 tons and managing on a tank of water, a distance of 48¾ miles.

When Manson rebuilt these with domed boilers he retained the Stirling cab. The tenders were the most modern on this line to date and gave the engines a businesslike yet handsome appearance.

The near failure of the Stirling 0-4-4T on the Greenock line presented a problem which had to be resolved promptly if the company was to build up, and hold, a satisfactory traffic in and out of the Clyde port in the face of growing Caledonian competition.

Smellie found the answer in the first of his two classes of 4-4-0s, the '119' class, or 'Wee Bogies' as they were called to distinguish them from the later '153' class or 'Big Bogies', the difference being in the size of the driving wheels.

In 1882 the first of the '119's came from the company's works at Kilmarnock. The cylinders of these engines were 18¼×26 in., the larger diameter than before setting a new standard. As these engines were specially wanted for the Greenock road the coupled wheels were 6 ft. 1¼ in. diameter. For every pound of available pressure there was 120 lb. of tractive effort and all of this would be needed on the climb out of Greenock to Upper Port Glasgow where the ruling gradient was 1 in 70.

The boilers were very similar to those of the 0–6–0 of the year before and furnished 964 sq. ft. of tube surface and 101 sq. ft. of firebox area with a 16 sq. ft. grate. The boiler pressure was still only 140 lb. per sq. in. The engines weighed 41 tons 10 cwt. of which 13 tons 15 cwt. rested on the bogie, and the tenders were similar to those of the '22' class goods engines. In the four years to 1885 twenty-four were built, all at Kilmarnock, and as they became more numerous they were put to work on many of the principal coast trains, and once the Stranraer turntable had been replaced by a larger one, to that port also, but it was on the Greenock line where they did their best work. When they first appeared these engines were the most powerful passenger engines in the country.

The 'Big Bogies', the '153' class of 4–4–0s did not make their debut until 1886 when the first of 20 engines, built at Kilmarnock between then and 1889, came off the works. Modelled on the '119' class the 'Big Bogies' had the same size cylinders but the coupled wheels were 6 ft. 9½ in. diameter. The working pressure was raised to 150 lb. per sq. in. and the heating surfaces were: tubes 1,092 sq. ft., firebox 106 sq. ft., total 1,198 sq. ft. The grate area 17·5 sq. ft., the largest yet on the South Western. Engine and tender weights were increased, the former by one ton to 42 tons 10 cwt. and the latter to 29 tons 12 cwt., the water capacity being increased to 2,500 gallons.

Four of these engines were later fitted with an extended smokebox in the form of a drum of slightly smaller diameter than the smokebox itself. This was the first application of this feature in Britain and only lasted until about 1895 when the longer smokebox became a more regular feature.

In August 1888 Queen Victoria visited the Glasgow Exhibition and was brought from Carlisle by the G. & S.W.R., the only reigning monarch to be conveyed by this company. The Royal Train was handed over at Carlisle two minutes late but a 'right time' arrival was effected by a good run behind Smellie's 'Big Bogie' No. 70;

this engine thereafter bore the royal arms on the cab side panels in commemoration of the occasion.

One of the class, No. 65, was so badly broken up in the Barassie accident of February 4, 1898, that it was never repaired. The remaining 19 engines all passed into L.M.S. hands. In their hey-day they were one of the fleetest 4-4-0 classes ever built in this country. Some were rebuilt by Manson with domed boilers, and like some of the 'Wee Bogies' came in for the attention of Whitelegg during his make-do-and-mend years after the First World War.

Whereas most of the Scottish companies used the customary balance weights for the reversing shaft Stirling had used a device like a watch spring. Smellie used an arrangement of two volute springs which he had found successful on the Maryport & Carlisle Railway. He also removed Stirling's safety valves from the tops of the fireboxes and placed them mid-way along the barrel thereby enhancing the appearance of the engines.

The steam brakes applied to the 0-6-0 goods engines were in the nature of an experiment. The engine only was fitted with a brake cylinder; it was not till the early 1900s that Manson extended the fitting of brake cylinders to the tenders. In operation steam was admitted above and below the brake piston so that the latter floated in the cylinder. The brake was applied by cutting off the steam supply to the underside of the piston, allowing the pressure of steam on the top of it to push the piston down in the cylinder and so apply the brake through the usual linkage. There was a great deal of steam leakage which caused a haze of steam which was not conducive to good visibility. Later Smellie modified the design, blanking off the steam supply to the underside of the piston. The admission of steam to the upper side of the piston only was then necessary for a brake application, with a return spring to assist in taking the brake 'off'.

Smellie's bogies were of the long wheelbase type, which by now was becoming accepted practice in this country, pivoted at the centre. The pivot pin passed through a block which could slide laterally in the centre casting, the movement to either side being controlled by springs so arranged that to whichever side the block was deflected both springs were compressed instead of one being compressed and the other unloaded as in the ordinary arrangements. The advantages of this were four-fold. Since one control spring neutralized the other in the central position, softer springs could be used; the travel of the sliding block was less jerky; the increase in the restoring force as deflection increased was less rapid; and moreover, if one spring broke, all control on one side of the bogie was not lost. The spindle carrying the control springs was extended through the bogie frame plates and adjustment of the compression could be made by means of a nut and lock-nut readily accessible from the outside. Any tendency to rolling at the front end of the engine was checked by side cushion slides. The bearing springs of the bogie were laminated plate springs, one over each axlebox, the inner spring pillars being connected by a compensating beam rather similar to Cowan's arrangement on the G.N.S.R., though in this case the assembly was inside the frame not outside as in Cowan's design. There is no doubt that these bogies contributed in no small measure to the excellent riding of these engines.

During the summer of 1889 Smellie's health had given cause for concern and he was advised to take an extended holiday on the sea. From the voyage he returned greatly benefited and when Drummond left St Rollox in search of greater things in the Antipodes Smellie was appointed in his place. He took office at St Rollox on September 1, 1890, and was shortly afterwards embroiled in the arduous struggle with the men involved in the Scottish railway strike which commenced a few days before Christmas of that year. The three southern companies were badly affected, and so far as the Caledonian was concerned Smellie did much to obtain a settlement. However before a satisfactory solution was reached his health again gave way. He died in a Bridge of Allan Hydropathic establishment suddenly on April 19, 1891, at the early age of 51 years. A strict disciplinarian, he was nevertheless a man of innate charm and kindly and courteous manner, besides being a most capable engineer. His profession was the poorer for his passing. During his twelve years at Kilmarnock the grim battle of continuous brakes had been fought and Smellie had done his part conceding neither one brake nor the other to be the better without adequate trial.

In the 80s the South Western's territory terminated at Girvan. Thence forward to Challoch Junction, 1¼ mile east of Dunragit on the Dumfries–Stranraer line, was the property of the Girvan & Portpatrick Junction Railway. The South Western had worked this line from its opening in 1877 until 1886 when the G. & P.J. took over the operation of the line itself. Then in 1887 a new company, the Ayrshire & Wigtownshire was formed which took over the G. & P.J. and reigned until in its turn it was absorbed by the G. & S.W.R. The management of the G. & P.J. was in the hands of W. T. Wheatley, son of Thomas Wheatley, ex locomotive engineer of the N.B.R. who had been lessee and manager of the Wigtownshire Railway until his death in 1883.

The locomotives of the G. & P.J. included three 4–4–0T bought second-hand from the North London Railway and which were scrapped at the 1892 takeover, and four 0–6–0 engines designed by W. T. Wheatley and built, two by Neilson & Co. in 1886, and two by The Clyde Locomotive Company in 1887. These four engines were remarkably good ones to find on a remote and struggling secondary line, and they were up-to-date engines as well. All four had cylinders 17 × 26 in. and 5 ft. 1½ in. wheels. The boilers contained 200 × 1¾ in. tubes, 966·9 sq. ft., the fireboxes added 98·3 sq. ft. while the grate area was 16·5 sq. ft. and the working pressure 150 lb. per sq. in. The engines weighed 37 tons and the tenders 29 tons and carried 2,200 gallons of water. Westinghouse brakes were fitted, operative on the tenders and trains only, the engines having steam brakes. Holt & Gresham's steam sanding apparatus, the invention of Fred Holt at the Derby Works of the Midland, was fitted to these engines. To assist in the negotiation of sharp curves water jets were provided in front of the leading engine, and behind the trailing tender, wheels as a form of flange lubrication.

During the first ten years of their lives these were hard-worked engines, mainly on the Stranraer–Glasgow boat trains between the southern port and Girvan. After the amalgamation with the G. & S.W. they reverted to goods trains and their places on passenger workings were taken by some of Smellie's 'Wee Bogies'.

In 1875 James Manson, a Kilmarnock trained man who had been engaged in marine engineering for several years, returned to the railway and was almost at once put in charge of the work of fitting the company's stock with Westinghouse brakes. When Smellie took office in 1878 he made Manson his general foreman; this was really the position of works superintendent, a post vacated by the removal of Robert Campbell to the Maryport & Carlisle in succession to Smellie. Then, in 1890, on Smellie's appointment to the Caledonian Railway the board gave Manson the post of locomotive superintendent. Manson was then at Kittybrewster. Born in 1845 at Saltcoats James Manson was still quite young when his family moved to Kilmarnock where in due course he followed his father into the railway service. His apprenticeship was mainly under Patrick Stirling and his later training and drawing office experience under James Stirling. In 1869 Manson left the railway to broaden his experience by entering the field of marine work both ashore and afloat. The eight years from 1875 until 1883 were spent on the South Western. In the latter year he succeeded Cowan on the G.N.S.R. and the first engines he designed for this company were the six 'A' class Nos. 63-8 which Kitson & Co. built in 1884. These engines were slightly smaller than Cowan's last three classes but were credited with being better steamers. The cylinders were $17\frac{1}{2} \times 26$ in. and the coupled wheels 6 ft. diameter. Though the firebox heating surface was a little greater at 90 sq. ft. the tube surface was considerably less, being only 964 sq. ft., and the grate area, 16 sq. ft., was a slight increase on Cowan's grates. The boiler was pressed to 140 lb. per sq. in. and the weights of engine and tender were 37 tons 2 cwt. and 29 tons.

Manson altered the appearance of G.N.S. engines by adopting a flush top firebox with the dome on the middle of three barrel rings instead of the raised firebox of the Clark-Cowan era with its superimposed large dome. Ramsbottom safety valves without any casing were mounted on the firebox. Perhaps the greatest difference was in the change from outside to inside cylinders.

Westinghouse brake equipment was fitted to these engines when they were delivered. The exhaust from the Westinghouse pump was directed into the ashpan. This tended to reduce clinker formation but increased the amount of ash. As in all G.N.S. engines until Johnson came, Clark's Patent Smoke Consuming Apparatus was provided. Two batches of these engines were delivered by the makers, the second having Manson's modification of Clark's fittings. Experience had shown that the steam jets were unnecessary when running with steam on, and when the regulator was closed the blower answered all requirements. The air tubes in two rows of eight, above the foundation ring, on the throat and door plates, were retained. The first three engines were also fitted with brick arches but these were omitted in the second batch.

In 1885 another three Kitson engines arrived. These were intended for goods or fish traffic. They had the usual $17\frac{1}{2} \times 26$ in. cylinders but the coupled wheels were reduced to 5 ft. 6 in. diameter. The boilers were similar to the 'A' class of 1884. The numbers of the new engines were 69, 70 and 71 and formed the 'G' class.

The years 1884–5 also the introduction of the only six-coupled engines ever owned by the Great North. Two classes of 0–6–0Ts were built by Kitsons, the 'D' class of six engines in 1884 and the 'E' class of three in 1885. These neat little engines had 16×24 in. cylinders and 4 ft. 6 in. wheels. The heating surfaces were: tubes 690 sq. ft., firebox 66 sq. ft. and grate area 15 sq. ft. The working pressure was 140 lb. per sq. in. The differences between the two classes were small. The boilers of the second group were pitched $\frac{3}{4}$ in. higher than the 'D' class and they were delivered with Westinghouse brakes fitted instead of steam brakes. The weights were $37\frac{1}{2}$ tons and $38\frac{1}{2}$ tons. It was expected that these tank engines would be put on shunting and banking duties but they were to be found on branch passenger and goods workings and on the Aberdeen suburban, the 'subby' services. One of the 'D' class engines was generally utilized on the Fraserburgh & St Combs line which, being unfenced, required the fitment of a cowcatcher front and rear of the engine. Even as late as 1951 under the aegis of British Railways the St Combs Light Railway was being operated by steam engines which were ex Great Eastern 2–4–2Ts fitted with cowcatchers.

When these 'D' and 'E' class engines first went to work they found all the weak spots in the permanent way and derailments, especially in sidings, were frequent. Until the necessary repairs could be taken in hand the extraordinary measure was taken of removing the leading sections of the coupling rods in order to permit these engines to run as 2–4–0s with a shorter rigid wheel-base. All the 0–6–0Ts were rebuilt by Pickersgill between 1907–11.

The small repair shop at Kittybrewster was by no means adequate for the construction of new engines yet Manson managed by some miracle of organization and management to build two engines there in 1887. These were Nos. 4 and 5 and were named after the estate of the chairman *Kinmundy* and after the deputy chairman *Thomas Adam*. They formed the 'N' class and the frames were supplied by Dübs & Co., the wheels and axles by Sharp, Stewart & Co. while Kitson & Co. supplied the boilers. The latter were larger than the 'G' class, having 1,088 sq. ft. of tube surface, 99 sq. ft. of firebox, a 16·5 sq. ft. grate and 140 lb. pressure. The coupled wheels were 5 ft. 7 in. diameter but the cylinders were similar to the 'G' class, $17\frac{1}{2} \times 26$ in. The engines scaled 40 tons and the tenders 29 tons.

Increasing traffic and accelerated services called for improved motive power and in 1888 Manson designed the 'O' class, and again the builders were Kitson & Co. Although it was claimed that between £300 and £400 had been saved on each of the home-built engines the experiment was not repeated.

In addition to increased dimensions the 'O' class incorporated several new ideas. The cylinders were the largest yet on the G.N.S., 18×26 in., the coupled wheels were 6 ft. $0\frac{1}{2}$ in., the tube surface was 1,059 sq. ft. but the fireboxes had 106 sq. ft. of surface and the grate area was 18·2 sq. ft. The working pressure remained 140 lb. per sq. in. These engines were 4 tons heavier than the 'N' class and the tenders weighed the usual 29 tons. An innovation was the introduction of the swing link type of bogie, now used for the first time in Britain. The swing links had double top pins and it was claimed that, with this arrangement, when the bogie was deflected the front end remained parallel with the rails instead of being tilted.

Hitherto in his designs Manson had continued the practice of placing the slide valves between the cylinders but in the 'O' class he placed them on top of the cylinders and partially balanced them by means of circular relieving rings held against the underside of the steam-chest covers by light tripod springs. Wear and tear of valves was reduced and operation of the reversing gear by the driver eased. Investigation into the amount of wear of slide valves of this pattern showed that after the first 100,000 miles the thickness of the valves had been reduced by only $\frac{3}{16}$ in.

The next two classes were the 'P' and 'Q' classes, of which three of each were built by Robert Stephenson & Co. in 1890. The first named, Nos. 12–14 were almost identical with the 'O' class, and the 'Q' class, Nos. 75–7, were similar but had 6 ft. $6\frac{1}{2}$ in. coupled wheels and were needed for the summer express traffic. Not arriving until August 1890 they were kept in store until the 1891 season began. Both classes had very large bogie wheels, 3 ft. 9½ in. diameter, and the fore end frames were set in to accommodate the swing of the bogie when deflected on curves.

The tenders were unique and, together with two similar tenders introduced by Manson on the G. & S.W.R. some years later, were the only examples in this country. They were eight-wheeled, the two leading axles were mounted in a bogie frame, and the two rear axles were rigid. This arrangement gave a slightly greater degree of flexibility to the otherwise rather long wheelbase.

Until Manson's coming, the G.N.S. had been an outside cylinder line with a legacy of the Allan front end as modified by Clark, and later Cowan. Manson would have none of this but developed a clean-cut design which, especially in the Kitson built engines, proclaimed a family likeness to the engines this firm was building for other lines. He also changed the style of the chimneys, scrapping the earlier copper caps and substituting a chimney rather similar to those of Patrick Stirling at Doncaster.

In July 1890 Manson obtained the position at Kilmarnock vacated by his brother-in-law Hugh Smellie, and took up his new duties there on September 1st. On the drawing-board at Kittybrewster he left a design for an 0-4-4T engine, but it was left to his successor James Johnson to modify and build it.

In addition to the locomotive designs Manson had produced in the six years he served the Great North, he was responsible for the development of an apparatus for the automatic exchange of tablets on single line sections. The appointment of William Moffatt as general manager was the first stage of the reorganization and general improvement of the service performed by the company. He at once began to revolutionize the train services, ably seconded by Reid and Manson. Such was the improvement that the 'O' class of 1888 was working over the Huntly–Aberdeen section in 42 minutes for the 40¾ miles. The generally higher standard of performance was however limited to some extent by the need for drastic reduction of speed in order that tablets might be exchanged on single line sections.

The oft-repeated story of Manson and the Kittybrewster Blacksmith collaborating to produce what became known as the

Manson Automatic Tablet Apparatus does not bear the light of close investigation. It appears that the idea originated from either (or both) Ferguson the chairman, and Moffatt. Nevertheless Manson was authorized to develop the necessary apparatus. He did so and refrained from patenting the device, if it was his own, either on the grounds that a safety appliance should be free for adoption by any other company, or because of the urgent promptings towards greater safety, both for passengers and staff, by the Board of Trade officers in their reports on accidents. The ordinary exchange of tablets at slow speeds, done by hand, was simple enough, but something of a hazard for signalmen and enginemen when attempted at speed.

As a result of its adoption on all single line sections of the G.N.S. in 1889 tablets could be exchanged at speeds up to 50 m.p.h. and 60 m.p.h. with consequent saving in time and fuel. Both the Highland and Caledonian companies took it up as did one of the Irish companies, but the G. & S.W.R. to which Manson returned in 1890 would have none of it.

Until Drummond's advent the Caledonian had been an outside-cylinder line, and had followed almost without a break, from its inception, the Allan or 'Old Crewe' pattern of fore end design. Now all this was to be changed and some thirty years were to pass before the company again built an outside cylinder locomotive. Drummond would countenance inside cylinders only, and the Allan style was swept away from all Scottish lines except the Highland.

Towards the end of Brittain's regime the directors had ordered an investigation into, and reorganization of these works and the department as a whole. This exercise entailed a very considerable enlargement of the works together with new machinery. Drummond's recommendations were made shortly after his arrival and his initial proposals authorized in October 1882, work commencing on the new shops soon after.

This work retarded the production of new locomotives and the expansion of the Caledonian system was such that larger and more powerful engines were urgently needed for goods and mineral trains, as also were passenger engines commensurate with increased loadings and higher speeds, and to eliminate the amount of double heading of trains which was on the increase. A further essential requirement was a considerable measure of standardization of component parts in an endeavour to cut down overhead charges by reducing the variety and number of materials held in the main stores and at running sheds.

Before Drummond started to build any engines to his own design he set about the rebuilding of the Conner '417' class 2–4–0. During the years 1883–6 twenty-five of these were rebuilt and given new boilers of his own design; the tube surface was reduced to 837·95 sq. ft., the firebox surface increased to 101·07 sq. ft. and the grate area to 18 sq. ft. This big increase in grate area, and therefore capacity to burn coal, together with the increased firebox volume and the boiler pressure raised to 150 lb. per sq. in., enabled these engines to handle the main line passenger traffic until such time as the new passenger engines should be built. a number of alterations in the appearance of the 'Rebuilds' as they were called, made them characteristically Drummond. The slotted splashers were filled in, the stove pipe chimneys replaced by Drummond's own design, a simple cab of the Stirling pattern substituted for Conner's and the safety valves, Ramsbottom type, were mounted on the domes.

When one considers the excellence of Drummond's earlier engines, certainly those built for the N.B.R. and Caledonian, it is rather surprising that he should have committed such a gaffe as to tinker with the Conner 8 ft. singles to their detriment. These engines were distinctive in having long travel valves and the excellence of their steam distribution was reflected in their running.

After Drummond's ministrations, with their valves reset with short laps and shortened travel, their performance was sadly affected, and they were never the same again.

Once the rebuilt goods engines were out of the way and out on their new workings, attention could be paid to the new goods engines which were needed. Modelled on his N.B. 0–6–0, the new engines were more powerful than any which had gone before. St Rollox being immersed in the process of enlargement and having the rebuilds to contend with, the first of the new class came from Neilson & Co. in 1883. The Springburn firm turned out 15 and St Rollox did manage to turn out four. The '294' class, better known as the 'Jumbos', had 18 × 26 in. cylinders, 5 ft. wheels and

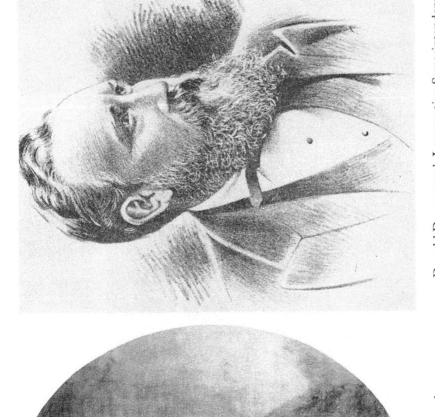

22. Dugald Drummond, Locomotive Superintendent, Caledonian Railway, 1882–1890. *Mitchell Library, Glasgow.*

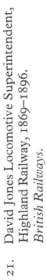

21. David Jones Locomotive Superintendent, Highland Railway, 1869–1896. *British Railways.*

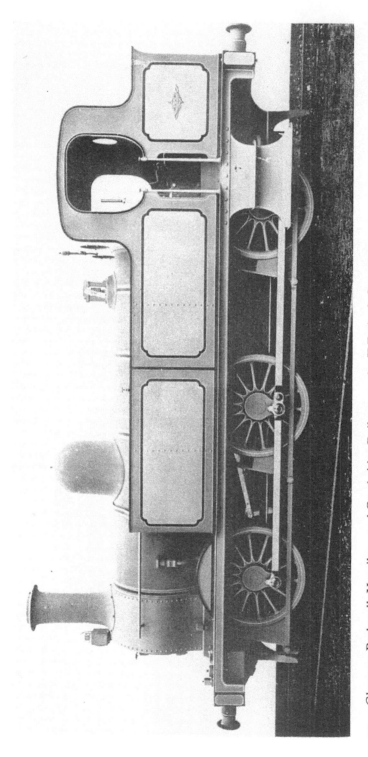

23. Glasgow, Bothwell, Hamilton and Coatbridge Railway, 0–6–0T Dübs & Co., 1877.
NBR Nos. 502–5.
Mitchell Library, Glasgow.

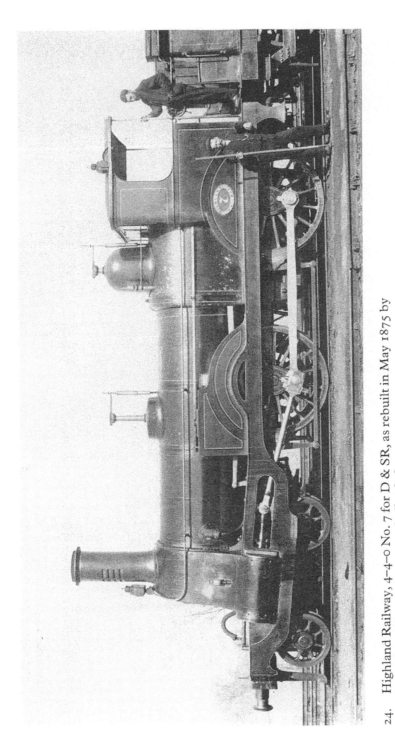

24. Highland Railway, 4–4–0 No. 7 for D & SR, as rebuilt in May 1875 by Jones from Hawthorn 2–4–0, built 1858. *British Railways*.

25. Caledonian Railway, 2–2–2T No. 1200, Glasgow, Paisley & Greenock R. 2–2–2 of Sinclair's 1846 design, as rebuilt 1881 and used to haul Officers' Saloon. *Mitchell Library, Glasgow.*

a working pressure of 150 lb. per sq. in. giving them a tractive effort of no less than 17,901 lb. at 85 per cent B.P. The tube surface was 1,089·68 sq. ft., firebox 112·62 sq. ft. and the grate area 19·5 sq. ft.

The arrangement of cylinders and valves was to become almost typical of Drummond. To reduce the effect of connecting rod angularity conical pistons were used, permitting a longer connecting rod. In a similar manner, by supporting the slide bars at the centre, longer eccentric rods could be used with consequent improvement in the steam distribution. The valves were placed vertically between the cylinders and the exhaust ports were divided so that part of the exhaust steam was directed round the cylinders in the form of an exhaust steam jacket, thus reducing the condensation of steam. The crank axles were forged solid wrought iron and the wheel centres were of Krupp's cast steel. Integral with the wheel centres were the balance weights which were confined to the space between three spokes of the drivers and two spokes of the leaders and trailers instead of the more usual crescent type spanning several spokes.

In 1875 Drummond patented his method of tyre fastening (patent No. 1986, May 31, 1875). In this arrangement the outer edge of the tyre was turned to form a lip and recess, and a projection was turned on the wheel rim which accurately fitted the recess. On the inside a groove was formed in the tyre and a similar groove in the rim. A ring of 'U' section was accurately fitted in the two grooves and was held in position by laying over a lip on the inside of the tyre. This means of fastening tyres was carried on as standard Caledonian practice for many years. McIntosh was using the same method after the turn of the century. It may well be the forerunner of the modern 'Gibson Ring' fastening.

What was to become known as the Drummond cab was first brought out on these engines and set a pattern for a number of years. The tenders were almost pure Stroudley though this was the last class on which Drummond used underhung springs.

At first the 0–6–0s were painted in the Caledonian blue livery and one depot, Motherwell, in addition to the vermilion panel on the buffer beam, painted the coupling rods vermilion. Fortunately this practice did not become general as it made the examination of the rods for flaws impossible unless the paint was first removed.

Ultimately there were 244 engines in the class. The earlier ones had steam brakes only but later some were fitted with Westinghouse brakes for working passenger trains, and in addition five were fitted with condensing apparatus to enable them to work on the Glasgow Central (underground) Railway.

In all, 35 were built by Neilson & Co. and 12 by the company up to the end of 1885; thereafter the company built the remainder of the class, 197 engines, spread over the years until 1897. During the first three years of the McIntosh regime 83 were constructed at St Rollox with some modifications, e.g. McIntosh supported the slide bars at their rear ends instead of at the middle as in the original design, and he used longitudinal girder stays on the firebox crown in place of the Drummond arrangement of four rows of sling stays at the front of the box and direct stays behind them. Fore end lubrication was improved by the substitution of Furness lubricators for the Drummond tallow cups, and the tender springs were over- instead of under-hung. McIntosh also followed the change Lambie had introduced by putting the safety valves on the firebox rather than on the dome top.

The same entry in the directors' minute book recording the order placed with Neilson & Co. for the first 15 'Jumbos' also recorded an order, with the same firm, for ten passenger engines and tenders. The day following the firm's tender for the 25 engines at £2,900 each was accepted but it was not until 1884 that the first 4-4-0 was delivered. Seven more came out in that year and the last two arrived in 1885.

Basically these engines were N.B. 'Abbotsford' 4-4-0s with a much improved boiler. The cylinders were 18×26 in. and the coupled wheels 6 ft. 6 in. As before, the coupling rods were 9 ft. 0 in. long. The tube heating surface was 1,088·7 sq. ft. made up of 1,047·2 sq. ft. from $216 \times 1\frac{3}{4}$ in. tubes and 41·5 sq. ft. from $10 \times 1\frac{1}{2}$ in. tubes. The length between tubeplates was 10 ft. 7 in., a Drummond standard, and the barrel was 10 ft. 3 in. made up of two rings only, the dome surmounted by the safety valves being set on the back ring. The grate area was 19·5 sq. ft. and the working pressure 160 lb. per sq. in. at 85 per cent, of which the tractive effort was 14,688 lb.

The first ten engines, Nos. 66–75, were contract built but their successors were the St Rollox products. Nos. 60–5 appeared in

1885, followed in 1889 by Nos. 76–9, 84 and 87, and in 1891 by 83, 88–91, and 113. A 29th engine was added to the class under rather special circumstances.

In the great Exhibition held in Edinburgh in 1886 engineering was to play a large part. The Glasgow locomotive builders all sent magnificent examples of their art, two of which were at the behest of the Caledonian. Mr A. G. Dunbar has been at pains to unravel the mystery surrounding the ordering of these two locomotives, a 4–4–0 from Dübs & Co. and a 4–2–2 from Neilson & Co. It is not intended to go into the details of the transaction here; it is enough to say that these two engines were built by the respective firms to Caledonian drawings, the directors' minute recording acceptance of the tenders at a cost of £2,600 each engine for 'exhibition purposes'.

The 4–4–0 was No. 124 and in most dimensions was exactly similar to the previous 28 engines save for the cylinders which were 19 in. diameter with the slide valves on top, and the valve gear which, instead of being the usual Stephenson link gear was the Bryce-Douglas gear. Archibald Bryce-Douglas was another Ayrshireman; he haled from Ardrossan, and was employed by the Fairfield Shipbuilding & Engineering Company. He had designed a valve gear mainly for marine engines. In the field of locomotive engineering the application was purely experimental. It owed something to Joy but also had a claim to having been derived in part from Walscheart, part of the valve spindle movement being taken from the crosshead. The reciprocating motion thus obtained was combined with the oscillating movement of the expansion link, derived through a complicated system of links from a connection with a lug on the underside of the connecting rod. The lifting links for the valve rod were connected to an arm of the reversing shaft in similar manner to the Stephenson link engines, the shaft being beneath the motion due to lack of head-room, the boiler centre line being pitched only 7 ft. 3 in. above rail level. For the high speed running required by an express locomotive there were too many pin joints, some, notably in the connecting rod, subjected to maximum stresses, and on several occasions No. 124 strewed her motion on the permanent way. The name *Eglinton* was bestowed on her later, and in general service with the orthodox link motion she did some very good work.

The second exhibition engine ordered by the Caledonian board was Neilson & Co's contribution, a very handsome bogie single numbered 123. The why and wherefore of this wheel arrangement are shrouded in mystery. That Drummond was the prime mover in the general characteristics cannot be denied. Both engines shout Drummond from buffer to buffer. The likeliest answer to the question 'who designed these engines?' is Snowball, Neilson's chief draughtsman, and his opposite number at St Rollox, William Weir.

The usual Drummond cylinders, 18×26 in., were set horizontally between the frames and drove a pair of 7 ft. driving wheels. The trailers were 4 ft. 6 in. diameter and carried the rear end of the engine 8 ft. behind the drivers. The bogie wheels were 3 ft. 6 in. on a 6 ft. 6 in. wheelbase, the centre being 9 ft. 10 in. forward of the driving axle and the total engine wheelbase being 21 ft. 1 in. The tender wheelbase was 13 ft. evenly spaced between the three axles and the capacity was $4\frac{1}{2}$ tons of coal and 2,850 gallons of water.

The boiler was 10 ft. 7 in. long and carried $196 \times 1\frac{3}{4}$ in. tubes giving 950·1 sq. ft. the firebox provided 103 sq. ft. in a total of 1,051·1 sq. ft. There were six longitudinal stays in the barrel and the firebox crown was supported by 12 rows of sling stays suspended from four angle-irons riveted to the wrapper plate. The grate area was 17·5 sq. ft. and the working pressure 150 lb. per sq. in.

The pistons were Drummond's conical type and the cylinders were steam jacketed. The blast pipe was of the Vortex type developed on similar lines to that of Adams on the L. & S.W.R.

The bogie bearing springs were laminated and the side control was by a pair of laminated springs whose centres bore on the central sliding block. Laminated springs were used for the trailing axleboxes but the drivers were given a pair of coil springs each.

As was typical of Drummond's practice at the time the boiler was fed by two injectors mounted alongside the ashpan and delivering through clack-boxes on the boiler centre line approximately 2 ft. behind the smokebox.

In 1905 a modified boiler was fitted, increasing the heating surface by 30 sq. ft., and the pressure was raised to 160 lb. per sq. in.

The weight distribution as built was: bogie 13 tons 17 cwt.,

driving axle 16 tons 19 cwt. 1 qr., and the trailing axle 11 tons
1 cwt. making a total of 41 tons 17 cwt. 1 qr. Despite the very
low adhesion factor of 2·75 this remarkable engine did some very
fine work, particularly during the race to Edinburgh in 1888. It
passed into L.M.S. stock in 1925, being renumbered 14010 and
painted in the lake livery of the new company. It was withdrawn
from service in 1935 after having run 780,000 miles, and is now in
the Glasgow Museum of Transport in her old Caledonian blue
livery amongst other preserved Scottish locomotives.

The Caledonian answer to Smellie's 'Wee Bogies' for the Greenock road was also a 4-4-0 but with coupled wheels only 5 ft. 9 in.
diameter. Standard 18 × 26 in. cylinders were used and the boiler
was somewhat smaller than that of the 6 ft. 6 in. engines. The
total heating surface was 936·5 sq. ft. with a grate area of 16·5
sq. ft. Twelve were built at St Rollox, Nos. 80–82 in 1887,
85, 86 and 116 in 1888 and 114, 115, 95–8 in 1891. They were an
instant success on the Greenock services which were extended to
Gourock on 1889 in the grim competitive battle for the Clyde
steamer traffic. The Caledonian route was far easier than that of
the G. & S.W.R. which traversed the heights of the Renfrewshire
hills, whereas the line followed by what was originally the Glasgow,
Paisley & Greenock Railway was almost level for most of the
distance, compared with the rival route.

After the success of his main line engines Drummond turned
his attention to suburban and branch line services, and for them
designed a tank engine which was a complete reversal of his N.B.
practice. The wheel arrangement he chose was 0-4-4 whereas his
previously successful N.B. engines had been 4-4-0s.

Now it is well known that tank locomotives are more subject to
weight variation than tender engines, as the contents of bunker and
tanks diminish. In practice there is little to choose in this respect
between the two types mentioned though it must be conceded that
the rearing couple exerted due to tractive pull at the drawbar,
causing an engine to sit down as it were, at the trailing end, will in
fact favour the 4-4-0 tank rather than the 0-4-4T. This may have
been the reason for Drummond's initial choice of the 4-4-0 type.
Whatever the reason may have been, 24 0-4-4T were built at
St Rollox 1884–91. With 16 × 22 in. cylinders, 5 ft. coupled wheels
and 150 lb. per sq. in. pressure, they had a tractive effort of

11,968 lb. The boilers were proportionately small, 3 ft. 10 in. diameter with $138 \times 1\frac{3}{4}$ in. tubes giving 607·3 sq. ft. The firebox added 65 sq. ft. and the grate area was 14 sq. ft. These engines were quite Stroudleyesque, particularly in the cabs and bunkers.

Following these branch line engines came some additions to the fleet of 0–4–0 saddle tanks for the Glasgow dockland. Based on the Neilson engines of 1875 these little engines were built at St Rollox in 1885 at a cost of £700 each. Small outside cylinders 14×20 in. drove on to 3 ft. 8 in. wheels. A boiler pressure of 140 lb. per sq. in. and the barrel 3 ft. $7\frac{3}{4}$ in. diameter was pitched only 5 ft. $4\frac{1}{2}$ in. above rail level. The tube surface was 632 sq. ft. and the firebox surface 52 sq. ft. in a total of 684 sq. ft. The grate area was 10·23 sq. ft. Eight of these engines were built though ten were authorized, but the same year saw two 0–4–2Ts also built at St Rollox to the same dimensions. These two engines were Nos. 262 and 263, the eight 0–4–0STs being Nos. 264–71. The 0–4–2Ts, which were also saddle tanks, went to the Killin Railway which had been opened on March 13, 1886. To enable them to work passenger trains on this steeply graded line, both engines were fitted with Westinghouse brakes. The tanks held 800 gallons of water and the bunkers 15 cwt. of coal. Later the coal capacity was increased to $1\frac{1}{4}$ tons by the fitment of coal rails to the bunkers.

Two classes of 0–6–0 saddle tanks were built. The first was in 1887, the year of Queen Victoria's Golden Jubilee, hence the class became known as the 'Jubilee Pugs'. Ten engines were built with 18×26 in. cylinders, 4 ft. 6 in. wheels, and they weighed 43 tons 17 cwt. The boilers contained $177 \times 1\frac{3}{4}$ in. tubes, 837·95 sq. ft., the fireboxes gave 101·07 sq. ft. and the grate area was 17 sq. ft. Coal and water capacities were 2 tons and 950 gallons respectively. The overall length was 31 ft. $2\frac{1}{2}$ in. with a wheelbase of 15 ft. 6 in. and a trailing overhang of no less than 8 ft. $10\frac{1}{4}$ in. In 1888 eight more appeared, followed by another twelve in 1890.

1888 also saw the production of six 'Dock Pugs', neat little 0–6–0 saddle tanks with 3 ft. 8 in. wheels and 14×20 in. cylinders. The boilers were very similar to those of the 0–4–0STs and Killin pugs.

This concludes Drummond's contribution to the motive power of the Caledonian Railway and it was by no means ordinary. His initial work on the neighbouring N.B.R. had paved the way for the very significant impact he made on the larger railway, significant in that he set a pattern which was followed, not slavishly but intelligently and with considerable success, by those who followed him on both companies. It must not be overlooked that basically his designs were the logical development of the Stroudley school as interpreted, and improved on, by him. He made his mistakes as do all men, witness the abortive 0–4–2T on the N.B.R. and the altered valve setting of the Conner singles, but he profited from those errors of judgement and, as the next decade was to show, he was as much in advance of his time as a later generation was to find Churchward in the south.

At the height of his success on the Caledonian he was attracted by what appeared to be a lucrative post in Australia and accordingly he terminated his service with the Caley in July 1890. He was succeeded at St Rollox by Hugh Smellie from the G. & S.W.R. at Kilmarnock.

When Matthew Holmes assumed control of N.B.R. locomotive fortunes in 1882 he had had seven years of Drummond's training, counsel and leadership. Now a man of 46 he had been his chief's chief inspector, virtually the assistant locomotive superintendent. His whole career had been spent in the Edinburgh area. He was born in Paisley in 1836, and while he was still a child his father entered the service of the Edinburgh & Glasgow Railway in the capital city. Young Matthew went to Hawthorn & Co., Leith, to serve his apprenticeship, then followed his father into the E. & G.R. In 1865 Wheatley made him locomotive foreman at Edinburgh, Haymarket, a post he held until Drummond promoted him in 1875. There has been a story several times told, that Holmes had been appointed in Paton's time to be in charge of the locomotive department of the Stirlingshire & Dunfermline Railway, a subsidiary of the E. & G. Indeed Bradshaw's Railway Shareholders' Manual for 1855 records an M. Holmes holding that post. However the M. Holmes with whom we are now concerned was then only 19 years of age and still serving his time at Hawthorn's.

On the departure of Drummond, Holmes did not, as so often happened, start to scrap all his predecessor's work and generally reorganize again. He was too well imbued with Drummond's teaching and had had too much of a hand in shaping the department under his chief's guidance, so it was not until 1884 that anything new came off Cowlairs shops other than orders initiated by Drummond.

Holmes's first engines were six comparatively small 4–4–0s not intended for express work but for secondary duties. Nos. 574–9 came from Cowlairs in 1884 and had the usual two cylinders 17×24 in., 6 ft. 6 in. coupled wheels and a pressure of 140 lb. per sq. in. The total heating surface was 1,059 sq. ft. and the grate area 17 sq. ft. The coupled wheelbase was 8 ft. 3 in., nine inches shorter than Drummond's engines. Holmes adopted the Stirling type of cab, similar to that employed by his predecessor on his rebuilds, and he did not perpetuate the use of Ramsbottom safety valves on the dome but used a lock-up valve in the same position. These features, cabs and safety valves, became standard Holmes practice.

Two years later a much improved 4–4–0 was brought out. This was the '592' class which numbered twelve engines all built at Cowlairs in 1886–7. The cylinders were increased in size to 18×26 in. and the coupled wheels to 7 ft. on the exceptionally long wheelbase of 9 ft. 3 in. This was necessary to get the size of grate required, 21 sq. ft., horizontally between the driving and trailing axles, Holmes preferring this arrangement to Drummond's sloping grate. The boilers were 4 ft. 4 in. diameter and contained 202×1¾ in. tubes (brass) giving a heating surface of 983 sq. ft. which, added to the firebox surface of 119 sq. ft., produced a total surface of 1,102 sq. ft. There were six longitudinal stays. The engine weight was 40 tons 15 cwt. whilst that of the tender was 32 tons with 5 tons of coal and 2,550 gallons of water.

Inverted laminated springs were used on the bogies with coil springs for the side control. There were laminated springs under the driving axleboxes and coil springs under the trailers.

Screw reversing gear mounted on a pedestal was fitted in the cab and Westinghouse brakes were applied to the coupled wheels, the brake blocks being on the rear of the drivers and the front of the trailers. The pump was mounted on the right side of the firebox. Injectors, Drummond fashion, were fitted at the sides of the ashpan which had a front damper only.

The first engine, No. 592, was shown at the Exhibition in Edinburgh in 1886.

In 1890 No. 602 was used to haul the special train conveying the Prince of Wales, later King Edward VII, at the opening of the

Forth Bridge. To commemorate the event the engine afterwards carried the Prince of Wales' feathers on the upper panels of the cab inside sheets.

Holmes built two classes of 0–6–0 goods engines. The first came from Cowlairs in 1883 and were modifications of Drummond's design with 17 in. cylinders. Further engines were built in 1885, 1886 and 1887 until thirty-six had been added to stock. These engines had boilers similar to the 574 class of 4–4–0. The cabs were Stirling type and the safety valves of Holmes's lock-up type. The underhung springs of the tenders were replaced by springs above the axleboxes.

The second class of goods engine appeared in 1888 and had 18 in. cylinders with 26 in. stroke. The wheels were 5 ft. diameter. Cowlairs built no less than 138 of these, Neilson & Co. fifteen, and Sharp, Stewart & Co. 15 in 1891 and 1892 respectively. The heating surface was 1235·13 sq. ft. and the grate area 17 sq. ft. with a working pressure of 140 lb. per sq. in. The final 24 had the pressure raised to 150 lb. per sq. in. The number of tubes varied; the early engines had $238 \times 1\frac{3}{4}$ in. tubes, later engines 236 tubes of the same size. The loss in tube surface was compensated by the firebox area which at 104·72 sq. ft. was large for the period. Between 1913–20 some rebuilding was carried out. The boilers were made larger, $2\frac{1}{4}$ in. on the diameter bringing it up to 4 ft. $8\frac{1}{8}$ in. with $252 \times 1\frac{3}{4}$ in. tubes, 1,214 sq. ft., while the fireboxes were actually reduced to 95 sq. ft. At the same time Reid, who was then in charge, resprung the engines with laminated springs under the leaders and trailers and helical springs under the drivers. Even later Chalmers put helical springs under all axleboxes but this arrangement proved too soft and there was a reversion to the Reid springing. Eighty-seven of the class had steam brakes, 13 steam and vacuum brakes, 48 had Westinghouse and 20 Westinghouse and vacuum brakes. Such versatility ensured they were available for almost any kind of traffic and to run with other company's stock, passenger or goods, and 25 served overseas during the first World War. On their return they were named after generals and Western Front battles.

These too were the engines usually used for the attachment of snow ploughs. The ploughs Holmes used were very like those Stroudley had designed for the Highland and were large wooden structures mounted on the fronts of the engines, bolted to the buffer beams and lifeguards. From the top of the plough a wire was taken back and made secure round the chimney, with a screw toggle to effect adjustment of the riding height of the plough above the rail.

Of passenger tanks also there were two types. In 1886 Cowlairs built six 0-4-4Ts, Nos. 586–591, with 17 × 24 in. cylinders and 5 ft. 9 in. coupled wheels. The boilers were again similar to the '574' class and the pressure was 140 lb. per sq. in. They had lever reverse and weighed 51 tons 4 cwt. When new these engines had Westinghouse brakes only but in 1913 two of them were fitted with vacuum brakes, by 1920 three more were dual fitted, the last engine being fitted after grouping. Rebuilding was carried out in 1911–13 when boilers similar to those of the Holmes 17 in. goods engines were fitted. Nos. 90–5 came out in 1888 to satisfy the need for more engines of the same type though this batch carried a working pressure of 150 lb. per sq. in. As it turned out these were the last passenger tank locomotives to be built by or for the N.B.R. for 21 years.

Holmes also built a number of dock pugs of the 0–4–0ST type between 1887–9. Altogether 34 were built, all at Cowlairs. In general they had 14 × 20 in. cylinders, 788 sq. ft. of heating surface and 130–40 lb. per sq. in. pressure. Dumb buffers, saddle tanks and inclined cylinders characterized these engines which were to be found in dock areas all over the system. In addition to these there were two Neilson 0–4–0 tanks built in 1882 and taken into stock by the N.B. in 1884.

In 1885 Holmes made his one essay in compounding. No new engine was chosen for this experiment but one with a history, No. 224, 'The Diver'. Considerable alteration was necessary. The original cylinders were replaced by two low-pressure cylinders 20 in. diameter; in advance of these were two high-pressure cylinders 13 in. diameter co-axial with the low-pressure cylinders and having a common stroke and one piston rod passing through both cylinders on the same side. The stroke, as on the original engine was 24 in. The valve gear was a modification of Joy's radial gear, arranged so that the high and low pressure valves could be notched up independently of each other. Of the four valve spindles the outer pair served the low-pressure valves and the inner pair the high-pressure valves. In order that the valve spindles could be thus arranged the high-pressure valve faces were machined at an angle of about 45 degrees to the horizontal. Both sets of valves were over the cylinders. From a T-joint near the top of the smokebox tubeplate the steam pipes were led by sweeping curves to the high-pressure steam chest which was well in front of the smokebox. The exhaust from the high-pressure cylinders was taken direct to the low-pressure steam chest through a short large diameter pipe.

The boiler was renewed at the same time and contained $164 \times 1\frac{3}{4}$ in. tubes and seven longitudinal stays. The firebox crown was stayed by means of girders slung from two double angle-irons.

In reporting on this engine the *Engineer* on October 22, 1886, said it had been working passenger trains between Edinburgh and Glasgow with great freedom and had shown excellent results in speed, power and fuel consumption, which latter showed a marked decrease compared with the simple engines. It is a great pity that no test data have ever been published for this interesting experiment; apparently, however, it did not fully realize Holmes's expectations, for, after a few years, it was re-converted to a two cylinder simple engine.

While these events were taking place in the south David Jones was having considerable success with the 4-4-0 engines he was putting into service on the Highland Railway, so much so that in 1886 a new series of eight engines was ordered, this time from the Clyde Locomotive Works. This company had been formed in 1884 when Walter Neilson had severed his connection with Neilson & Co. some years previously, leaving James Reid in control at the Hyde Park Works. The new venture was by no means a success, in fact the first locomotives built in the new works, across the line from the Hyde Park establishment, were the eight Highlandmen.

The first engine was delivered in May 1886. This was No. 77, the firm's No. 2. The remainder appeared at intervals during the year, the last to be delivered being No. 76, works No. 1; the whole class bore names as was the Highland custom, No. 76 being *Bruce* and giving the name to the class. This was the engine chosen to represent the builders at the Edinburgh Exhibition and its presence there from April 1st accounts for its non-delivery at Inverness until December, when the Exhibition closed.

Jones was sufficiently progressive to have adopted a good degree of standardization, only varying essentials where they were relevant to the class of work for which the engine was required. As in his previous classes the cylinders of the new engines were outside the frames and were cast separately from the steam-chests, and, like the '60' class, were 18 × 24 in.; similarly the coupled wheels were again 6 ft. 3 in. diameter. The valve gear was the standard Allan straight link motion, the valves having 1 in. lap, $\frac{1}{4}$ in. lead, and 4 in. travel in full gear. The maximum cut-off was at 75 per cent and the minimum at 15 per cent. Frames and wheels were as in the '60s'.

The boilers were shorter than those of the '60s' and Skye Bogies by nearly 1 ft. but contained the same number of tubes, 223 × $1\frac{3}{4}$ in. The heating surfaces were: tubes 1,038 sq. ft., firebox 102 sq. ft. a total of 1,140 sq. ft. and the grate area was 18·83 sq. ft. The working pressure was 160 lb. per sq. in.

Whereas the '60' class had Adams patent pop safety valves the new engines had $2\frac{1}{2}$ in. lock-up valves. In addition to the Le Chatelier counter pressure brake the engine and tender were fitted with the Automatic Vacuum brake. The bogie had independent laminated springs over each axlebox and spiral spring side control. Laminated springs were also used for the coupled wheels and these were compensated. The tender also had laminated springs, over the axleboxes, and each of the three axles had inside and outside bearings. The engines weighed 43 tons, 28 tons of which were available for adhesion, and the tenders weighed $31\frac{1}{2}$ tons with 2,250 gallons of water and 5 tons of coal. The class was mostly withdrawn by the end of 1923 though one engine ran until 1930.

Two further classes of 4–4–0 were to come from Jones during his superintendency, the 'Strath' class in 1892 and the 'Loch' class in 1896. The former were built by Neilson & Co. and the latter by Dübs & Co.

The 12 'Straths' were all delivered in the months of May and June. Cylinders, frames, motions and wheels were all as in the '60' and '76' classes but the boilers were larger to accommodate 242 tubes and were therefore 4 ft. 5 in. diameter. The $1\frac{1}{4}$ in. red metal tubes gave a heating surface of 1,127 sq. ft., the firebox 115 sq. ft., a total of 1,242 sq. ft. The grate area was again 18·83 sq. ft. as in the Clyde bogies. Gresham & Craven injectors were fitted on these engines, the first use of this type of injector in Scotland. The tenders were similar to those of the 'Bruce' class and the engines weighed 45 tons, distributed as follows: bogie 15½ tons, driving axle 15 tons, and trailing axle 14½ tons.

During and after World War I some repairs of a heavy nature were contracted out to private firms and No. 81, *Colville* of the '76' class, was rebuilt by Robert Stephenson & Co. in 1916. Some of the 'Straths' were given steel fireboxes by North British Locomotive Co., Hawthorn Leslie & Co. and William Beardmore & Co. of Dalmuir. A number of this class had been scrapped by 1923 though some remained until 1930.

The last of Jones's 4–4–0s were very fine engines turned out by Dübs & Co. in 1896 incorporating some altered features of design. The cylinders were increased to 19 in. diameter, the stroke remaining 24 in. Piston valves as designed by W. M. Smith of the N.E.R. were provided. A sloping grate replaced the hitherto standard horizontal grate and, as in the case of the 4–6–0 goods of two years before, the Allan front end was conspicuous by its absence, though wings were retained on the smokebox front plate.

The coupled wheels were the standard 6 ft. 3 in. on a 9 ft. wheelbase, and the bogie wheels 3 ft. 3 in. on a 6 ft. 6 in. wheelbase, the total engine wheelbase being 23 ft. 6 in. On this rested the 49 tons of the engine, 16 tons 1 cwt. on the drivers and 15 tons 9 cwt. on the trailers. The boiler diameter rose to 4 ft. 6 in. and it had 244 × 1¾ in. tubes at 10 ft. 6¼ in. giving 1,176 sq. ft. heating surface whilst the firebox contributed 119 sq. ft. to the total of 1,295 sq. ft. The long sloping grate was 20·5 sq. ft. in area and the working pressure was 175 lb. per sq. in. The tender water capacity was raised to 3,000 gallons and with 5 tons of coal the weight was 38 tons 11 cwt.

The piston valves were not particularly successful at this stage of development and during Peter Drummond's period in office they were replaced by Richardson balanced slide valves. Not enough was known about the lubrication of piston valves even though the additional problems caused by the use of superheated steam were still unknown in the 1890s. It is likely that some difficulty was caused by the high proportion of downhill running on the Highland line, when steam would be almost, if not entirely, shut off and the valves lacking proper lubrication.

In September 1896 the Duke of York was conveyed from Perth to Grantown-on-Spey and thereafter the Prince of Wales' feathers were borne on the driving splashers of No. 119, the engine used to draw the special train.

In all 15 were built in 1896 and of these ten were rebuilt with Caledonian type boilers and boiler fittings in 1924–8 under L.M.S. auspices. During World War I when the Highland was suffering from serious shortage of power three identical engines were built by N.B. Loco. Co. at their Queen's Park works. These three engines were fitted with both Vacuum and Westinghouse brakes.

In 1890 Jones brought out his first tank locomotive, a saddle tank 0–4–4 No. 13 built at Inverness. It had 14 × 20 in. cylinders, 4 ft. 3 in. coupled and 2 ft. 7½ in. bogie wheels. The boiler, which came off one of the 1862 Hawthorn singles No. 13 withdrawn at the same time, had 578 sq. ft. of heating surface (tubes), having been considerably shortened to serve its purpose in this engine. The firebox was reduced to 62 sq. ft. and the grate to 12·5 sq. ft. The tank saddled the boiler immediately behind the smokebox and held 820 gallons of water. Between the rear of the tank and the front of the cab, one the firebox top, a very large dome with Salter safety valves was mounted. The bunker carried 1½ tons of coal and the working pressure was 100 lb. per sq. in. As with all Highland engines at this time it had the Jones louvred chimney. In 1900 it was renumbered 53 and in the following year was again rebuilt with a new Drummond boiler and with side tanks instead of a saddle tank. The heating surfaces were improved to 623 sq. ft. tube surface, 67·5 sq. ft. firebox and the grate to 13 sq. ft. The pressure was raised to 140 lb. per sq. in. and, whilst the bunker was unaltered, another 80 gallons of water was carried in the side tanks. The effect of rebuilding was to increase the weight from 32 tons to 34 tons with 19½ tons on the coupled wheels. This engine was withdrawn by the L.M.S. in 1929.

Dübs & Co. constructed some 4–4–0T engines for the Uruguay Eastern Railway in 1891 but for some reason the American company did not take delivery. Two were purchased by the Highland on a year's trial and were given the numbers 101 and 102.

The cylinders were 16 × 22 in. and the coupled wheels 5 ft. 3 in. diameter. The boilers were a little larger than the 1890 engine, the diameter being 3 ft. 10 in. and the length of the barrel 10 ft. 6 in.

Heating surfaces were: tubes 795 sq. ft., firebox 88 sq. ft., total 883 sq. ft., and the grate area was 14 sq. ft. The pressure was 140 lb. per sq. in. and the engines weighed 42·5 tons. The valves were over the cylinders and driven by rocking levers. The slide bars were covered in when the engines were delivered, presumably as some protection from dust and sand on their intended Uruguayan line. When newly arrived at Inverness they were put to work on the Burghead branch on trial. After twelve months' satisfactory service they were accepted into Highland stock and three identical engines were ordered from Dübs & Co. and were delivered late in 1893, being numbered 11, 14 and 15. Three of the five 'Yankees' as they were called, Nos. 102, 14 and 15, were later named. The L.M.S. withdrew one in 1924, one each in 1927 and 1928, and two in 1934, the latter being the first two delivered.

At this period of Highland Railway history there did not exist a single six-coupled locomotive for main line work. Other Scottish companies were using them, except the G.N.S., and had not Drummond introduced the 'Jumbo' to the Caledonian? But the Highland preferred the 4-4-0 for goods work as well as passenger work. Certainly an 0-6-0 could not have handled the regular traffic unaided, let alone the seasonal tourist trains on which they would have had to play a part, moreover if one considers the terrain over which the line was constructed it is probable that Authority felt that a leading guiding truck was essential for safe running at high speeds down the many gradients.

Thus when the need arose for larger and stronger engines to cope with the growing traffics, and the opening of the Forth Bridge in 1890 had its effect here, Jones introduced a radical change from preconceived ideas of British locomotive designed. In 1894 he produced a design for a six-coupled goods locomotive with a leading bogie and in February of that year a contract was placed with Sharp, Stewart & Co. This firm had succeeded The Clyde Locomotive Co. and had taken over the latter's works. The Manchester works of Sharp, Stewart had been called the 'Atlas Works' and this name was transferred to their 'new' premises in Glasgow. The order was for 15 engines and tenders, not just a prototype to be tested first but the complete order of 15 without any preliminaries. The first five engines were delivered in September, seven in October and the remaining three in November.

They were the first 4–6–0s to be supplied to any British railway. True, there had been some of the type built for overseas railways but these were the first for use at Home.

The cylinders, 20 × 26 in., were outside and were not embraced in the Allan fore end arrangement of the frames, the first departure from this tradition in 40 years. Coupled wheels nominally 5 ft. 3½ in. diameter and a working pressure of 170 lb. per sq. in., later raised to 175 lb. per sq. in. produced a tractive effort of 24,362 lb. not much less than their successors, the ubiquitous 'class 5' of the Stanier era of L.M.S. management.

The boilers of these new engines were of steel, the first such on the Highland, and were 4 ft. 7⅞ in. diameter with 211 × 2 in. tubes. The heating surfaces were: tubes 1,559 sq. ft., firebox 113·5 sq. ft. total 1,672·5 sq. ft. with a grate area of 22·6 sq. ft. The firebox was 7 ft. 9 in. long outside and 7 ft. 0½ in. inside. The front set of firebars was on a slope, the rear set being horizontal.

The bogie wheels were 3 ft. 2½ in. diameter on a 6 ft. 6 in. wheelbase. The coupled wheelbase was 5 ft. 6 in. leaders to drivers, 7 ft. 9 in. drivers to trailers, and the total 25 ft. The bogie carried 14 tons, the leading coupled wheels 13 tons 10 cwt., drivers 14 tons 10 cwt. and the trailers 14 tons making a total weight of 56 tons. The tenders were six-wheeled carrying 5 tons of coal and 3,000 gallons of water and weighed 38 tons 7 cwt.

The valve motion was the traditional Allan gear. Two engines of the class were fitted with Westinghouse brakes, for handling foreign stock, in addition to the standard vacuum brake which was fitted to the whole class. These engines were allotted the numbers 103–17 and were not named. Spark arresters were fitted to the bases of the chimneys which were of the Jones louvred pattern.

Arriving too late for the tourist season of 1894 they were all well run in by the time the 1895 season commenced and from the start were eminently successful engines, handling both goods and passenger trains in a manner that could not have been bettered, despite the low factor of adhesion; with 42 tons on the coupled wheels this figure was no more than 3·83, but this does not seem to have been detrimental. No. 103 has been preserved in the Glasgow Museum of Transport.

In 1870 Kitson & Co. of Leeds built a locomotive for the Third Duke of Sutherland who had his own small standard gauge railway

26. Glasgow & South Western Railway, Old 126–126A, Smellie 'Wee Bogie', built 1882.
Rebuilt by Whitelegg, 1921.
By courtesy of D. L. Smith.

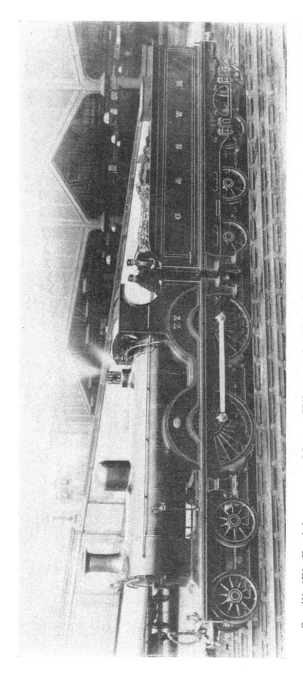

27. Smellie 'Big Bogie' 153 4–4–0 No. 77, Kilmarnock 1886–1889.
     As rebuilt by Mauson.
     D. L. Smith's collection.

28. Highland Railway, Jones Goods 4–6–0 No. 103 Class, 1894.
By courtesy of Photomac Ltd.

29. Great North of Scotland Railway, steam rail motor No. 29/28, 1905. *Andrew Barclay Sons & Co. Ltd.*

at the family seat, Dunrobin, which gave its name to the engine. It was an outside cylinder 2–4–0T and had been used on the Golspie-Helmsdale section of the Sutherland Railway, known as the Duke's Railway since he had subscribed the major portion of the cost of construction.

This little engine had 10 × 18 in. cylinders, 4 ft. coupled wheels, 379 sq. ft. heating surface and a 6¼ sq. ft. grate. The boiler pressure was 140 lb. per sq. in. and the weight, with ½ ton of coal and 400 gallons of water, 21 tons. The Highland acquired this engine in 1895 and arranged for Sharp, Stewart & Co. to overhaul and rebuild it. On its return to Inverness it had 12 × 18 in. cylinders, 522·4 sq. ft. heating surface and 8·4 sq. ft. of grate. The coal and water capacities were improved to ¾ ton and 560 gallons. Ramsbottom safety valves replaced the original Naylor valves. The engine was renumbered 118 and named *Gordon Castle*.

Shortly after the delivery of the first 4–6–0 engines David Jones was out riding on one of them, and it is said that he had the load of one train augmented from 49 wagons to 60 and this big load was successfully taken up the 'Hill'. On one such occasion he suffered a severe scalding of a leg and although he was spared the loss of the limb his health was so undermined that he resigned his post as from October 31, 1896.

In this chapter the endeavour has been to show how the three original trends, those of Stirling, Allan and the Stroudley/Drummond complex, developed during the first sixty years from their Scottish origins, and in some instances even blended.

Perhaps the most telling effect of the evolution of the steam locomotive was not noticeable to the travelling public until the prominence given by the press of the country to the races to Edinburgh in 1888 and Aberdeen in 1895. In the former the only Scottish company directly involved was the Caledonian in partnership with the L.N.W.R. since the East Coast workings were confined to the two English partners in the triumvirate, the N.B.R. having no hand in the Anglo-Scottish running. It is true that the Drummond/Neilson single No. 123 achieved glory on the Carlisle-Edinburgh run night after night, but in the races of 1895 it was Drummond's splendid Caledonian 4–4–0s and their Holmes cousins on the N.B. that really made the sparks fly.

Reference has several times been made to the financial difficulties of the Scottish companies and the measures often necessary to make ends meet so far as the locomotive stocks were concerned. Tight purse strings meant economy and an avoidance of waste, and the care exercised in order to carry on their business is reflected in some of the figures published in *Engineering* on October 26, 1886, showing three English and three Scottish companies' operating costs. The companies were all running into Carlisle, and the period referred to is the first half of 1886. The appended table shows that the Scottish companies were all more cheaply worked than their English contemporaries. An eighth column has been added to indicate the cost of locomotive working, repairs and renewals per train mile as a percentage of the traffic receipts.

For the whole year 1885 the Caledonian with 690 engines ran a total 17,692 miles per engine, the receipts were 4s 7d per train mile and the locomotive running and repair costs 4·54d and 2·49d per train mile respectively. The N.B. figures were 585 engines, 20,092 miles per engine, receipts per train mile 4s 4¼d, locomotive costs were 4·6d and 1·59d respectively. These results showed that the Caledonian locomotive costs were 12·78 per cent of the receipts against the N.B. figure of 11·4 per cent.

Comparative Loco Costs Railways Running to Carlisle
1st Half Year 1886

Company	Number of locos	Train miles per engine	Receipts per train mile	Cost of loco working	Cost of loco repairs and renewals		Loco costs % of receipts
					per engine	per mile	
L. & N.W.R.	2323	7848	5s 1¾d	5·44d	£84 9s 9d	2·58d	12·98
Midland	1750	9324	4s 4¼d	5·05d	£85 15s 4d	2·21d	13·88
North Eastern	1506	7371	4s 10¼d	5·81d	£132 12s 4d	4·35d	17·74
Caledonian	690	8777	4s 6¼d	4·52d	£95 0s 6d	2·60d	13·12
North British	591	10184	4s 2¾d	4·51d	£68 4s 11d	1·61d	12·09
G. & S.W.R.	291	8846	4s 6¾d	4·26d	£76 3s 6d	2·07d	11·56

It can hardly be said that money was thrown about and yet the travelling public was very well catered for and the railways of the country stood well as commercial enterprises. In no small measure was this due to the excellence of the locomotive designs produced by their engineers who knew their craft intimately and plied it courageously.

Chapter VII

JAMES MANSON LEAVES HIS MARK

After the retirement of David Jones from the Highland Railway in 1896 the Allan influence was at an end in Scotland except for a few isolated examples built as a make-up measure during the engine power shortage in World War I. On the Great North of Scotland this trend had completely vanished with the introduction of Manson's engines in which he incorporated a nice blend of Kilmarnock practice with the lines of the Kitson family of Leeds-built locomotives. The Caledonian had long since lost the Allan influence when Dugald Drummond introduced his interpretation of the Stroudley school of design.

The latter was in full bloom on the North British Railway where Holmes was developing the theme Drummond had brought from Brighton and worked on and improved, whilst on the Glasgow & South Western Railway the Stirling tradition was still evident though due shortly for an overlay of the Kitson element as brought from Kittybrewster by Manson.

As will presently be seen, the Stroudley/Drummond theme was to be carried on on the Caledonian by Lambie, McIntosh, and lastly by Pickersgill with ever larger, heavier and necessarily more powerful locomotives until the amalgamations of 1923.

In 1890 the new superintendent on the Great North was James Johnson. He was the son of Samuel Johnson of the Midland Railway, had trained at Derby under his father and had remained there to assist him. He arrived at Kittybrewster at a time when, as was really quite normal, the Great North of Scotland was at loggerheads with the Highland. These two companies were in a state of perpetual enmity, in fact the birth of each of them was conceived out of the animosity existing between their promoters.

The Great North had its eyes firmly fixed on Inverness, headquarters of the rival company, so-called capital of the Highlands, and an attractive prize withal. The wrangles over interchange of traffic centred on Keith and Elgin, the Highland insisting on the former as the exchange point and incidentally gaining the benefit of greater mileage thereby, gave rise to the establishment of a seven year traffic agreement in 1886. Despite this, disputes continued, and when it terminated in 1893 matters were as disturbed as ever.

The uncertainty of the outcome of the company's attempts to increase its field of endeavour (all of which is clearly told by Barclay-Harvey, Vallance and Nock), had a depressive effect on the locomotive department so that it was not until 1893 that any new engines were constructed for the Great North.

In that year Neilson & Co. built the nine 0–4–4 Tank engines that Manson has designed before leaving the company. Johnson modified the arrangement and the number of tubes and fitted his own type of dome, safety valves and chimneys; a strong Derby appearance was found in these features. The safety valves were duplicated, one set encased in the shapely brass column on the firebox crown, the other set on the dome but minus the Salter balances. This arrangement did not suit Pickersgill when he followed Johnson in 1894 and Ramsbottom valves were substituted, on the firebox.

These 0–4–4Ts were somewhat similar to the Glasgow & South Western engines built in the same year by Neilson & Co. to Manson's requirements, but apart from Johnson's alterations there were some differences.

The cylinders were $17\frac{1}{2} \times 26$ in. and the coupled wheels 5 ft. 1 in. diameter. The heating surfaces were: tubes 1,094 sq. ft., firebox 113 sq. ft., total 1,207 sq. ft., and the grate area was 18 sq. ft. All these dimensions were greater than the South Western engine save that the coupled wheels were 1 in. smaller. Only the working pressure was the same, 165 lb. per sq. in. The weight was $53\frac{1}{2}$ tons.

The only express engines built during Johnson's short span were the six 'S' class 4–4–0s of 1893. Again Neilson & Co. were the builders and in these engines it would seem that Johnson took Manson's later ones as the basis and superimposed his own ideas of contemporary Midland practice. The boilers were identical with the 'R' class 0–4–4Ts and the cylinders and coupled wheels were 18×26 in. and 6 ft. 1 in. respectively. These engines were reported to be capable of rapid acceleration and sustained high speed.

Johnson discontinued the use of Clark's smoke consuming apparatus despite the fact that it accounted for a saving in fuel consumption of around 2 lb. per mile.

In one important particular Johnson's practice differed from Manson's. The motion drove direct on to the valve spindles instead of through rocking levers. This required the valves being placed between the cylinders instead of on top of them. In a general way the 'S' class became the standard on which future designs were based. In August 1894 Johnson resigned to take up an appointment with a firm of engineers in the West of England and was succeeded by William Pickersgill.

Entering the service of the Great Eastern Railway in 1876 (he was born at Crewe 15 years previously), Pickersgill served his apprenticeship at Stratford and during this period he became a Whitworth Exhibitioner. In 1883 he was made an inspector in charge of tests with compounds and oil fired locomotives under Holden, and eight years later became district locomotive superintendent at Norwich. It was this post he now vacated to assume his new responsibilities at Kittybrewster and his arrival coincided with the height of the row with the Highland.

Since the Great North board did not lack faith that it would reach Inverness it was determined to put new engines into service to cope with what was hoped would be a build-up of north-west bound traffic. These hopes were short lived. Running powers over the Highland line were not authorized, and moreover, the 1895 race to Aberdeen had the result of speeding up services on the West Coast route though the G.N.S. did benefit to some extent from the earlier arrival of trains at Aberdeen. Nevertheless new engines were needed and Pickersgill's first contribution was a class of 26 engines all built by Neilson & Co. or their successors Neilson Reid & Co. between 1895 and 1898.

These 'T' class engines were 4–4–0s dimensionally similar to Johnson's differing only in such details as safety valves, etc. Ramsbottom valves were used instead of the Johnson arrangement, the tenders, except those of the first ten engines, had coal rails and from 1897 onwards the side rods were fluted instead of being of plain rectangular section. Nos. 93–9 were built in 1895, 19–24 in 1896, 101–6 in 1897 and 107–112 in 1898. During the period 1916–23 all these engines were rebuilt but none was superheated.

The next addition to the locomotive stock was a class of five very handsome 4–4–0s built by Neilson Reid & Co. in 1899. Originally the order was for ten engines and tenders and ten were built, but when delivery was being made there was a decline in traffic resulting in restrictions of some services. At the same time the South Eastern & Chatham Committee was urgently requiring some new locomotives. Accordingly the second five engines were sold to the English company which, although offered all ten, could only accept five.

In these engines there were several departures from previous G.N.S. practice, notably in the styling of the splashers and cabs. The latter had two windows on either side and clerestory roofs. As previously, the splashers swept from the cab front over the trailing and driving wheels in pleasing curves, but in the new engines their width was increased to embrace the sweep of the crank pins thus doing away with the need for the small secondary splashers hitherto necessary for this purpose. Ramsbottom safety valves, round topped dome, and shapely chimney, together with a balanced design of tender, contributed to their fine appearance. These locomotives formed the 'V' class and had the same principal dimensions as the two previous classes but the engines were one ton heavier. Thirteen were built altogether, over the years until 1915, the remaining eight, Nos. 27–9, 31, 33–6 were constructed in the company's own shops at Inverurie whence the department had moved in 1902. The first five were numbered 25, 26, and 113–5.

The early years of the twentieth century witnessed an attempt by a number of railways in Britain to reduce the operational costs of branch line and local services by the introduction of steam rail motors on some lines. The Great North and the Glasgow & South Western were the only Scottish companies to experiment in this form of transport and in 1905 two steam rail motors, to Pickersgill's designs, appeared on the Lossiemouth and St Combs branches of the former.

These units consisted of a small engine mounted on a four-wheeled bogie and a coach body, the front end of which was pivoted on the rear end of the engine bogie and whose rear end ran on an ordinary coach bogie.

The engine portions were built by Andrew Barclay & Co. of Kilmarnock, and were powered by vertical boilers made by Coch-

ran & Co. of Annan. These boilers were something new in locomotive work. They were adaptions of a type well known in other
fields but proved to be inadequate in this application. The lower
part of the boiler was cylindrical and contained the grate and tubes
and the upper part was hemispherical and contained the regulator
valve. Since the boiler was mounted vertically on the engine bogie
it gave the impression of a very large dome between the chimney
and the cab. Firing was done from the footplate in the usual
manner, onto a circular grate 9 sq. ft. in area, the firebars being
arranged as in crane boilers. The products of combustion were
brought away from the firebox at the side and deflected by a brick
arch through the lower half of a nest of tubes. The gases then
passed through the upper tubes, thence by a flue to the chimney.
The tubes, of which there were $295 \times 1\frac{1}{2}$ in. were arranged transversely to the longitudinal centre line of the engine and an inspection door, to facilitate tube cleaning, was provided on the sides of
the boiler casing.

The shell of the boiler was pressed out of a single flat plate and
was entirely seamless, with no rivets or welded joints exposed to
the action of the fire. Openings for firing and ash removal were
likewise without riveted flanges and the foundation ring was also
pressed from a single plate. This method of construction was used
to avoid localized heating at thick joints and the tendency to
grooving of the plates in the vicinity of the joints.

The diameter of the boiler was 6 ft. and the heating surface
500 sq. ft. The working pressure was 150 lb. per sq. in. Two $2\frac{1}{2}$ in.
Crosby type safety valves were fitted, one on either side in front of
the cab. A grid type regulator valve was mounted in the steam
space and controlled the flow of steam to the steam chests through
3 in. bore copper pipes. The 10×16 in. cylinders were just ahead
of the rear bogie wheels and drove on to the leading axle. A very
neat and compact Walschearts valve gear controlled the valve
events, the slide valves being set over the cylinders horizontally.
The maximum travel was $3\frac{3}{4}$ in., the lap $\frac{11}{16}$ in. and the lead $\frac{1}{8}$ in.
The wheels of the power unit were 3 ft. 7 in. diameter.

A small bunker attached to the front of the coach body formed
the back of the cab and held 15 cwt. of coal. Underneath the leading end of the coach there was a 650-gallon water tank. The frames
of the coach extended forwards and were carried on the engine

bogie frame. The central pivot was 8 in. diameter and rested in a cup in the centre casting secured by a 2 in. pin with a nut and cotter underneath. Due to the disproportionate weights resting on the bogie wheels, the pivot casting was positioned 6 ft. 8 in. behind the leading axle and 3 ft. 4 in. in front of the trailing axle. In consequence the springing of these axles was by laminated springs of different ratings. Three 9-coil bolster springs were provided for the leading bolster of the coach body.

The coach portion of the rail motor consisted of a long compartment for the accommodation of passengers. A small compartment at the rear end, with doors for ingress and egress of passengers, also served as a driving compartment when the unit was being driven from that end. To enable the driver to exercise control when driving from the rear end of the vehicle a ship's telegraph made by Messrs Chadburn of Liverpool and Glasgow was installed, a system of rods and levers operating the regulator valve, whilst the Westinghouse brake fittings were duplicated. The passenger compartment was 34 ft. 7 in. long and seated 45 while the overall length of the car was 49 ft. 11½ in. and the total weight 47 tons.

The two engine units were numbered 29 and 31, Barclay's numbers 1056–7. The coaches were Nos. 28 and 29. Unit 29/28 went to work on the St Combs Light Railway on November 1, 1905, and 31/29 started working on the Lossiemouth branch on the same day.

Although on test the rail motors showed good prospects they were a failure in service. There was nothing wrong with their speed, one attained a speed of 30 m.p.h. in 20 seconds and ran at 60 m.p.h. between Aberdeen and Inverurie. They were little more successful on the Aberdeen suburban services and in the course of time the engine units were detached from the coaches and used as stationary boilers. Here they were apparently more successful; on the line they were dreadfully noisy and the boilers would not steam properly, and the hopes of their designer were not realized.

During the 20 years Pickersgill served the G.N.S. he designed 39 locomotives and the two rail cars. Where Manson had changed the line from one featuring outside cylinders to one using inside cylinders instead, and Johnson had swept away the Clark smoke consuming gear, Pickersgill had introduced several changes as well. His cabs were roomy and really afforded protection to the enginemen. The side windows and clerestory roofs became a standard perpetuated by Heywood who succeeded Pickersgill in 1914.

Pickersgill removed all the old displacement lubricators, mostly of the Roscoe pattern, and substituted hydrostatic lubricators claiming that by their use the lubrication of valves and pistons was more efficiently achieved, a finer film of oil and less carbonization resulting.

Some criticism might be levelled at Pickersgill and his two immediate predecessors regarding the size of grate. This was never larger than 18·2 sq. ft. However the steaming capacities of the boilers was such that seldom was there any difficulty in keeping the cylinders fed, and as there was no non-stop runs of any great length the boilers were not put to any great strain.

James Manson returned to Kilmarnock in 1890. At this time there was an order for 20 goods engines of the 0–6–0 type in hand at Dübs & Co. To all intents and purposes they were similar to Smellie's '22' class of 1881. Manson immediately made several alterations to the design, raised the working pressure to 150 lb. per sq. in. and, strangely enough for a Kilmarnock-trained man, fitted domes to the boilers, although one of the class, No. 325, carried a domeless boiler for some years.

The cylinders were 18×26 in. and the wheels 5 ft. 1½ in. The boilers were dimensionally the same as those of the 'Big Bogies' and, at 39 tons 10 cwt., were 2 tons heavier than the '22' class. Vacuum brakes were fitted and these engines became very popular and were generally known as the 'Dübs Goods'. The L.M.S. rebuilt three of them with the X2 boiler of Whitelegg's design in 1924–5.

This, the '306' class, was the first of four developments of the 0–6–0 from this designer, and the first one came out in 1892. Five years later Kilmarnock built the 18 engines of the '160' class. These were the first goods engines wholly of Manson's design and they followed the previous engines very closely, almost the only variations being in the chimney and cab. The latter was rather higher than before and, since the boiler centre line remained at 6 ft. 10 in. above rail level, it gave these engines a hump-backed appearance. Until they were superseded by the '361' class, these engines were regularly employed on the 'Long Road Goods', the nightly goods train from Glasgow (College) to Carlisle via Dalry and Kilmarnock, rather than the more direct route via Barrhead. Three of the class got Whitelegg X2 boilers in the same manner as the '306s'.

In 1900 Neilson Reid & Co. built the first 20 of the '361' class. These were followed by 12 more in 1907, from the same works, though by this time Neilson Reid & Co., Sharp, Stewart & Co. and Dübs & Co. had amalgamated to form the North British Locomotive Company.

Again the cylinders were the same as their forerunners', as were the wheels, but there were slight differences in the boilers. The tube surface was 1,097 sq. ft., the firebox 111 sq. ft. and the grate area was 18 sq. ft. The working pressure was raised to 165 lb. per sq. in. and the engine weight to 44 tons. The greater length of barrel necessary to obtain these proportions was achieved by lengthening the wheelbase from 15 ft. 9 in. to 16 ft. 10 in. The boiler centre line was pitched 7 ft. 3 in. above rail and had the effect of radically improving the appearance of these engines.

Two additional engines of the class were built at Kilmarnock in 1910. Although to the same general design, these two engines were given the higher cab and larger tender of the '17' class which had then gone into production. By this variation some of the handsome outline of the other thirty-two engines was lost.

The '17' class came out in 1910, 15 being built by North British Locomotive Co. Once again the cylinders and wheels were standard with the previous classes but the boilers showed a big improvement. The tube surface was increased to 1,286 sq. ft., the firebox to 114 sq. ft. and the grate to 18·2 sq. ft. The working pressure was raised to 170 lb. per sq. in. These alterations increased the weight of the engine to 46 tons 14 cwt. and the big tenders weighed 34 tons.

Like each successive development of the goods engines they were put on the 'Long Road Goods' where they performed in excellent style with a load of 38 wagons, the maximum laid down for this class.

When he took charge at Kilmarnock, Manson's first undertaking was the completion of the 'Dübs Goods'. At the same time he was preparing the design for a passenger engine based on the G.N.S. 'P' class. The result was a fine looking and equally efficient engine which proved highly successful on the Coast and South main line services though without the facility for high speed possessed by Smellie's 'Big Bogies'. And yet no less than 57 of these 4–4–0s were built at Kilmarnock from 1892 to 1904. Generally known as 'Manson 4–4–0s' despite the fact that he designed four other classes of the same wheel arrangement, the '8' class had the usual $18\frac{1}{4} \times 26$ in. cylinders and 6 ft. $9\frac{1}{2}$ in. coupled wheels. In these engines Manson used his well-tried indirect motion for the valves. The usual Stephenson link gear, with crossed eccentric rods, drove a dummy valve spindle which worked in a sleeve bush in the spectacle plate. The forward end of this spindle was connected to the lower end of a rocking arm by a pin and die block working in a vertical slot in the rocker, and the upper end of the rocker was connected to an intermediate spindle in turn attached to the valve spindle proper, the slide valves being over the cylinders as in G.N.S. practice. The maximum travel of the valves was $4\frac{7}{16}$ in., the lap 1 in. and the lead $\frac{1}{16}$ in. Well-designed ports gave good admission and exhaust of steam, the steam admission ports being $1\frac{1}{2} \times 13\frac{1}{2}$ in. in area and the exhaust ports $5 \times 13\frac{1}{2}$ in.

The boiler barrel was 10 ft. 6 in. long and 4 ft. 3 in. diameter and contained $238 \times 1\frac{5}{8}$ in. tubes (brass) providing 1,094 sq. ft. of surface. The firebox surface was 111 sq. ft. and the grate area 18 sq. ft. A working pressure of 150 lb. per sq. in. produced a tractive force of 13,547 lb. Cast steel girder stays supported the crown of the copper firebox which was 5 ft. $6\frac{5}{8}$ in. long at the foundation ring and had a slope of 7 in. in a length of grate. The weights were 14 tons 4 cwt. on the bogie, 16 tons 10 cwt. on the drivers and 14 tons 7 cwt. on the trailers. To the total weight of 45 tons 1 cwt. was added the weight of the tender, 30 tons, the total for engine and tender being 75 tons 1 cwt. The tender carried 2,500 gallons of water and 3 tons of coal.

Two engines of the class had attached to them the eight-wheeled tenders of similar pattern to those Manson had introduced on the G.N.S. and was now bringing to the G. & S.W. These tenders were of 3,200 gallons capacity and were very necessary on the non-stop run between Carlisle and Glasgow, and showed to advantage on the 91·1 mile non-stop run from Carlisle to Kilmarnock when the Dumfries stop was made conditional. Such a run called for some extreme care in the consumption of water, allowing no more than 35 gallons to the mile, while on the full 115·5 miles non-stop no greater use of water than $27\frac{1}{2}$ gallons per mile was possible without an out-of-course stop.

The latter performance was on an even tighter rein than the Drummond 4-4-0 on the Caledonian, whose 3,572 gallon tenders allowed about 30 gallons a mile for the run from Carlisle to Stirling in the Race of 1895. The '8' class was economical on coal as well, the consumption on the 'Diner', the 1.30 p.m. ex Glasgow St Enoch, averaging $32\frac{1}{2}$ lb. per mile including lighting up, and this train loaded equal to $10\frac{1}{2}$, or around 150 tons.

The eight-wheel tenders were passed on from engine to engine as changes in allocation were made, and even from class to class as the '240' and '18' classes took over these jobs in their turn.

The next to appear on the scene was the very successful '336' class with 6 ft. $1\frac{1}{4}$ in. wheels specially for the Greenock road. Services on this line were becoming more and more exacting as competition with the Caledonian increased. In many respects these engines were very much the same as the '8' class but with 1,062 sq. ft. of tube surface. The firebox and grate were the same. The working pressure, at 165 lb. per sq. in., was the highest yet on this railway.

The weights of engine and tender were substantially the same as the '8' class.

For some years it has been stated by some writers that there were two distinct coupled wheel sizes in this class of South Western engine. It has been repeatedly claimed that the 6 ft. $1\frac{1}{4}$ in. engines were the first batch, Nos. 336-345, built by Dübs & Co. in 1895, and that the second batch of 15, Nos. 346-360 built by the same firm in 1899, had 5 ft. 9 in. coupled wheels. Such was not the case. The maker's order book, now in the custody of the Mitchell Library, Glasgow, shows that both orders were identical; in fact, against the second order, E.3731, is the note, 'Same as E.3239'.

These engines pressed to 165 lb. per sq. in. were the first on the G. & S.W. to have longitudinal stays in the boilers, a feature introduced by Manson for all engines with the pressure above 150 lb. per sq. in. One engine, No. 344, was fitted with a Drummond type of water-tube firebox in 1904. This arrangement consisted of two nests of tubes transversely across the firebox. Usually one set of tubes was inclined downwards from left to right and the other set inclined in the opposite direction. These tubes were almost twice the diameter of the boiler tubes and were of course of thicker gauge, the ratio of length to bore being 25 to 1 or better. The tubes were expanded into the firebox plates and were beaded over. The wrapper plate had to be cut out and a cover plate secured to seatings made of well annealed cast steel; by this means access to the cross tubes was afforded for inspection and cleaning. It is unfortunate that no records of this experiment are extant since it significantly augmented the heating surface of the boiler.

For some years these engines had a virtual monopoly of the Greenock traffic, in particular the 10.05 a.m. from St Enoch timed to reach Greenock in 37 minutes with a stop at Paisley (Gilmour St), and the 4.03 p.m. which was a 32 minute non-stop train. Both trains were advertised as 'No Luggage' in order to facilitate the rapid transfer of passengers from train to boat. Dunoon could be reached in 57 minutes by these trains and the connecting steamers. In the reverse direction the best time in 1900 was 40 minutes with a Paisley stop. This required some very fine work, the climb to Upper Port Glasgow requiring an output of something like 670 equivalent drawbar horsepower.

In 1897 Manson produced his only multi-cylinder engine. He was not seeking merely additional power but a more specific programme was set which can be summarized in the following main points:

1. To obtain a means of balancing the reciprocating masses by other means than the customary use of balance weights in the wheel rim.
2. To reduce the wear of big end brasses and overcome the well-known difficulty of providing adequate lubrication of big ends on long continuous runs.
3. To get increased crank area and subdivision of the greater power obtainable with higher pressures without complicating the machinery unnecessarily.
4. To relieve, or reduce, the knocking to which the main bearings are subjected when high pressure steam is admitted alternately to the opposite ends of a large cylinder.

To achieve these objects engine No. 11 was designed and turned out from Kilmarnock works in April 1897 and was the first British four-cylinder locomotive, forestalling Crewe by two months.

The outside cylinders were $12\frac{1}{2} \times 24$ in. and the inside cylinders $14\frac{1}{2} \times 26$ in. The coupled wheels were the standard 6 ft. $9\frac{1}{2}$ in. and the working pressure was 165 lb. per sq. in. so that the tractive effort at 85 per cent BP was 15,860 lb.

Slide valves were used for all cylinders, those for the outside cylinders being on top and fitted with the usual Manson spring loaded relieving ring, whilst the inside cylinders were served by valves in the more orthodox position between the cylinders. Only two sets of Stephenson link motion were required, being coupled to intermediate valve spindles driving the valves for the inside cylinders. These intermediate spindles were provided with pins and dies for the rocking arms which were connected to the spindles for the outside valves. The steam lap of the valves was $1\frac{13}{16}$ in. and the exhaust lap $\frac{3}{16}$ in. thus providing the characteristics which ensured a free-running locomotive when notched up. Despite this it seems that No. 11 was often off beat and no amount of tinkering with the valve setting was of any avail.

All the pistons drove the leading axle, the cranks being arranged in opposition so that the right inside was opposite to the right out-side, and the left inside opposite to the left outside. By this means the reciprocating masses were balanced by each other and hammer blow was eliminated and the objects of the first proviso were attained. Points 2, 3 and 4 followed from the first.

The bogie was of the swing link type which Manson stan-dardized on the G. & S.W.R.

The boiler was identical with that of the '8' class save for one feature. As in the 'Greenock Bogies' the pressure was 165 lb. per sq. in., seven of the boiler tubes were replaced by longitudinal stays. From time to time there have been misleading statements regarding the boiler details of this engine. These probably stem from the drawing which appeared in *Engineering*, August 10, 1900, which shows the general arrangement of the engine. The boiler is shown as having 238 tubes instead of 231. The tube heating surface was 1,062 sq. ft. not 1,094 sq. ft. as shown in the drawing. It seems likely that after the drawings had been sent to the journal for pre-paration of the lithograph Manson decided that because he had raised the pressure from 150 lb. to 165 lb. per sq. in. he needed increased strength in the boiler only to be got by substituting stays for tubes. This would account for the apparent discrepancy. The firebox and grate were the same as in the earlier engines and there were the usual fittings, Stirling steam reverser, G. & C. 40 mm combination ejector, two no. 8 injectors and a sight feed displacement lubricator in the cab fed into the main steam pipe. The cylinders were lubricated by Furness lubricators in the usual manner. The regulator was the standard double beat type and two 3 in. Ramsbottom safety valves were mounted on the firebox top. The total weight of the engine was 48 tons 10 cwt., of which 16 tons 16 cwt. was on the bogie, 16 tons 10 cwt. on the drivers and 15 tons 4 cwt. on the trailers. The tender weighed 32 tons 5 cwt. and carried 4 tons of coal and 2,500 gallons of water.

The life of this engine was very nearly terminated at an early age rather precipitately. On January 19, 1898, when only nine months old, No. 11 was attached to the down Pullman, then timed to leave Carlisle at 7.35 p.m., and at Border Union Junction ran into some derailed goods wagons of the 1.45 p.m. goods from Glasgow (College) which had been shunting across the road to leave the up road clear for the Up Mail. No. 11's bogie was found under the third coach of the Pullman and the engine was so badly damaged that it was doubtful whether it should be repaired or scrapped. In the event the former course was taken and No. 11 came out from the works a second time in March 1898.

Up to the date of the accident No. 11 had run 26,714 miles and in the subsequent 28 months to the end of July 1900 it had run a further 114,297 miles. These figures show that prior to the smash it was running about 120 miles per day, and in the period after March 1898 the figure rose to around 170 miles per day. This was in the days before 'intensive locomotive user' and 'cyclic diagramming of locomotives' were introduced to harass locomotive foremen.

In 1915 Peter Drummond put a '240' class boiler on this engine leaving the machinery as designed by Manson, but Whitelegg took it in hand in 1922 and completely reconstructed it.

The Mansons had proved particularly useful machines but growing traffics and increased loadings required something stronger yet smaller in size than the 4-6-0s of 1903. Thus in 1904-5 Manson built 15 engines of the '240' class at Kilmarnock in which, although the main dimensions were somewhat akin to the 1892 engines, the boilers were larger, the heating surface being increased to 1,315 sq. ft. obtained from 288×1⅝ in. tubes and a firebox of 119 sq. ft. The grate area remained 18·2 sq. ft. but the working pressure was raised to 170 lb. per sq. in.

This class did some excellent work on the coast and one was allocated to the 'Irishman', the boat train from Glasgow to Stranraer. Two of the class fell heirs to the eight-wheel tenders from later engines of the '8' class.

The last of Manson's 4–4–0s was the '18' class of which 13 were built at Kilmarnock between 1907 and 1912. Basically they were '240s' but with a long shallow firebox compared with the square and deep fireboxes of the earlier engines. The grate area was 22 sq. ft. About this time the L. & N.W.R. engineers at Crewe were conducting some experiments comparing the relative efficiencies of deep and shallow fireboxes with 'Precursors' and 'Experiments', and it may well be that Manson was influenced by them. Unfortunately no record has been left of the efficiency with which the '18' class performed their duties.

Two of them, Nos. 18 and 26, with the eight-wheel tenders, worked the non-stop Carlisle-Glasgow express which left St Pancras at 11.30 a.m. and was due at St Enoch at 8 p.m. Between 1912 and 1919 No. 27 was fitted with a Weir feed pump and water heater.

Although the G. & S.W.R. was finding that the 4–4–0 type was a good all round economical investment the pattern began to change when McIntosh brought out his huge 4–6–0s early in 1903.

At the same time train weights were increasing and a higher power output was being demanded from the locomotives. The Caledonian challenge was met by Manson who, on the instructions of his chairman, Patrick Caird, prepared a design for a 4–6–0. The North British Locomotive Co. was given the order for ten engines and tenders, Nos. 381–390, and they were built during 1903.

Two outside cylinders 20×26 in. drove the middle coupled axle. The coupled wheels were 6 ft. 6 in. diameter. The bogie wheelbase was 6 ft. 6 in., coupled 15 ft., and total engine wheelbase 27 ft. 8 in. The tender was carried on two four-wheel bogies each of 5 ft. 6 in. wheelbase at 11 ft. 6 in. centres and the total wheelbase of engine and tender was 55 ft. 5⅛ in. and the overall length over buffers 64 ft. 6 in.

The main frames were of 1¼ in. plate and were well braced. In addition to a cross stretcher in the normal position for an inside cylinder engine there was a second cross stretcher between the leading and driving wheels. Both stretchers were pushed to allow the intermediate valve spindles to pass from the quadrant block to the lower ends of the rocking shafts which in turn operated balanced slide valves situated horizontally over the cylinders. The negotiation of the leading axle was effected by an inset, horseshoe shaped, in the intermediate valve spindle so that the latter could operate clear of the axle. A screw adjustment was provided to assist valve setting. The valves had a steam lap of 1 in. and the throw of the eccentrics was $5\frac{1}{2}$ in. The angle of advance was $69\frac{1}{4}°$ and, since the drive was indirect the eccentric rods were crossed.

The boiler barrel, which was 4 ft. 8 in. diameter, contained 209×2 in. tubes giving 1,721 sq. ft. of surface at 15 ft. $8\frac{13}{16}$ in. between tubeplates, and the firebox contributed 131 sq. ft. in a total of 1,852 sq. ft. The grate area was 24·5 sq. ft. and had a slope downwards of 1 ft. 6 in. in a length of 8 ft. The tractive effort amounted to 20,400 lb. at 85 per cent of the 180 lb. working pressure. The engines weighed 67 tons 2 cwt., of which the bogie took 17 tons 5 cwt. while the coupled wheels bore 15 tons 4 cwt., 18 tons 1 cwt., and 16 tons 12 cwt. The tenders weighed 50 tons 6 cwt. and carried 4 tons of coal and 4,100 gallons of water.

The fireboxes of the '381' class were the first Belpaire fireboxes to be used in Scotland. The crown was stayed by four rows of sling stays suspended from two angle irons and there were sixteen rows of direct stays. The side stays were of Stone's bronze.

All wheels used laminated springs and the coupled wheel axleboxes were fitted with adjusting wedges. The bogie was given $5\frac{1}{2}$ in. side-play.

This class proved very efficient in service, reducing the amount of piloting necessary to a minimum, in fact only the heaviest of the Anglo–Scottish expresses required pilots for some years after their advent. The hard work they were called upon to do did however find some weak spots. The fore end framing was weak and caused trouble with the loosening of rivets and cylinder and stretcher bolts. In addition there was trouble from smokeboxes drawing air and from tubes leaking.

In spite of these ailments seven additional engines were built in the company's own works in 1910. The only significant difference between the first and second series was in the brake rigging and in the tenders. The 1903 engines had the brake hangers and blocks on the front of the leading and trailing wheels and on the back of the drivers, whereas the second lot were braked in front of all coupled wheels.

The last three engines of the initial series had attached to them tenders with six wheels instead of bogies like the other seven. The same kind of six-wheel tenders were attached to the second batch. This modification resulted in a saving in weight of 6 tons.

In 1920, under Whitelegg's superintendency, three were rebuilt with new boilers which had extended smokeboxes, not drum fashion as in the case of Smellie's engines, but well designed smokeboxes of adequate capacity. The grates were 7 in. longer allowing for an additional $2\frac{1}{4}$ sq. ft. area. At the same time larger roomier cabs were fitted. The remainder of the class got extended smokeboxes and, for a time, variable blast pipe nozzles were fitted.

During the whole of Manson's twenty-one years in charge of the locomotive fortunes of the G. & S.W.R. he only designed four classes of tank engine.

It must not be forgotten that tank engines were disliked on this line. In its whole history from 1840 until 1923 some 860 locomotives were constructed by or for the company or acquired by amalgamation. Of this total only about 108, or $12\frac{1}{2}$ per cent, were tank engines. Neither Patrick Stirling nor his predecessors had produced any, nor had Hugh Smellie. James Stirling had brought out two types and had rebuilt three tender types as tanks. Manson, Peter Drummond and Whitelegg produced seventy-four tank engines at least. Manson's contribution was only four classes of which one was an 0-6-0, one an 0-4-0 and two were 0-4-4; and when they were completed, in 1909, of 503 engines owned by the company only 46 were tanks, a mere 9·1 per cent. Of these only fourteen were available for working passenger trains.

The first of Manson's tanks to appear was a class of ten, the '326' class, which Neilson & Co. built in 1893. Although for the Glasgow suburban traffic they saw little of this work for a reason which was becoming common in other large centres of population. The growth of tramways on the city streets, with routes pushing ever outwards as the populace moved out from the business areas and the suburban areas grew, gave the public almost a door-to-door service between their homes and places of work. In consequence of this the rail services in the immediate environs of the city suffered from lack of patronage in much the same way as has happened in recent years as the internal combustion engine, in buses and private cars, has caused the withdrawal of many services operated by diesel rail-cars.

Thus some of the '326' class were put in store at Hurlford and brought out for use as and when required, but in the Glasgow area based on St Enoch they did do quite a lot of good work. With two $17\frac{1}{4} \times 24$ in. cylinders, 5 ft. 2 in. coupled wheels and a pressure of 150 lb. per sq. in., they were useful units on the local services which operated round the city.

The boilers were the usual Manson $238 \times 1\frac{5}{8}$ in. tube boilers with 1,056·24 sq. ft. of surface, the firebox gave 105·16 sq. ft. and and the area of the grate was 16·43 sq. ft. The tubes were of brass and the firebox the universal copper. Eight girder stays supported the crown. These engines weighed 51 tons 8 cwt. in working order. Two sets each of thirty-six bars furnished the grate and, strangely, the Clark smoke consuming apparatus, as modified on the G.N.S.R. by Manson, was fitted, there being 8×3 in. air tubes at the front and back of the firebox. The Author is indebted to Mr D. L. Smith for the information that the specification dated Kilmarnock 1893 states: 'The air tubes to be of best copper, solid drawn, 3 in. outside diameter No. 7 BWG thick, fixed in the firebox and casing plates with tube expander'. A pencil note adds: 'In one engine only, Mr Manson. April 7/1893.' No record of the behaviour of the fitting has been left.

The second of the 0–4–4T designs, the '266' class, did not appear until 1906; in that year Kilmarnock built six. These were smaller engines than the 1893 class and were intended for work in localities where sharp curves were prevalent. They commenced work on the Moniaive branch, the Cairn Valley Light Railway; the Catrine branch; and the Maidens & Dunure Light Railway. Ultimately they were concentrated at Ayr and Ardrossan where their usefulness was exploited in harbour shunting. The dimensions of the '266s' were very similar to those of the '14' class 0–6–0Ts which came out in 1896.

Manson had proved the utility of this wheel arrangement for shunting in goods yards on the Great North and in defiance of the

South West's dislike of tank engines produced this design, which subsequently numbered 15 engines. The initial four were followed by eleven others in 1903 and 1914. These two classes had 16×22 in. cylinders, 4 ft. 7½ in. coupled wheels, 820 sq. ft. of heating surface (tubes), 70 sq. ft. firebox, and 12 sq. ft. grates. In each case the working pressure was 140 lb. per sq. in. The weight of the '14s' was 40 tons 2 cwt. and of the '266s' 42 tons 8 cwt.

Whereas the bogies of the '326' class were inside framed, with the usual inverted laminated springs on each side, the smaller engines had a rather unsightly outside framed bogie with independent laminated springs over the axleboxes, similar to the Cowan 'H' class on the G.N.S.R. When Whitelegg rebuilt five of the 0-6-0Ts he extended the tanks to the front of the smokebox, enlarged the bunker and replaced the chimney by shorter ones.

The final Manson tank design was a masterpiece of compactness. *Multum in parvo* is an appropriate description of these little '272' class 0-4-0Ts, of which Kilmarnock built six in 1907-9. Harbour areas are notorious for abounding in curves which are little better than right-angle bends. Greenock harbour was no exception. Not only were there sharp curves to be negotiated but there were gradients of 1 in 50. The necessary power was obtained from 16×24 in. cylinders, 4 ft. 7½ in. wheels and 140 lb. per sq. in. pressure. The boilers were very like the two preceding classes and the weight at 39 tons 12 cwt. was almost double the weight per foot run of the '266' class since the wheelbase was no more than 7 ft. 6 in. The cylinders were placed outside the frames with balanced valves on top. Stephenson link motion was fitted between the frames and operated the valves through the unusual arrangement of rocking shafts in front of the cylinders. Hand brake and steam brake were provided.

The last of Manson's designs to be considered before the introduction of superheating is the steam rail motors of 1904-5.

Only three were built and they were superior to the two units on the G.N.S. not only in length of service but in performance as well.

The overall length of the G. & S.W.R. vehicle was 60 ft. 8 in. and this embraced a passenger carriage with seating for 50 passengers, and an engine unit. The carriage body was mounted on frames which extended beyond one end and between which the engine was positioned after the manner of a horse in the shafts of a cart.

The engine was a four-wheels-coupled well tank with two out-side cylinders 9 × 15 in. The wheels were 3 ft. 6 in. diameter. The boiler was of the locomotive type and contained 138 × 1¾ in. tubes giving 400 sq. ft. of surface, while the firebox gave 40 sq. ft. and the grate area was 8 sq. ft. Steam at 180 lb. per sq. in. pressure operated through slide valves on top of the cylinders, the valve events being controlled by Stephenson link gear between the frames. A small bunker carried 15 cwt. of coal and a 500 gallon tank was fitted between the frames; balanced safety valves were fitted on the dome. The wheelbase of the engine was 8 ft. as was that of the coach bogie, the two being at 39 ft. 4 in. centres.

The coach body was divided into three parts, that at the rear end being for the guard and for the driver when driving from that end. The other two compartments were for the use of passengers. Sliding doors gave intercommunication and side doors gave the means of ingress and egress. In addition, steps which could be lowered by a lever allowed passengers to enter or leave at road level when not at a platform. The seats were upholstered in plaited rattan cane and the floor covering was inlaid linoleum, the roof panelling was Alhambrine finished in white picked out with gold. Pintch's gas lighting was installed and parcel nets were fitted over the quarter lights. Hand and vacuum brakes were operable from either end, i.e. from footplate or from guards compartment, as was the whistle. Communication between driver and guard was by electric bell. The engines were painted in the standard crimson lake livery of the coaching stock instead of the G. & S.W.R. green.

When the Catrine branch was opened in 1904 one of these rail motors was allocated to the work on this branch and, such good steamers did the boilers prove, the whole day's work could be accomplished on two barrow-loads of coal. Another unit worked from Ardrossan to Largs and to Kilwinning, and the third went to the Moniaive branch.

There were several drawbacks to these units which were very soon found by the enginemen. The cabs were draughty; there was a constant splash over from the tank filler hole and the footplates were thus in a perpetual state of wetness so that the men heartily disliked the breed.

On the Moniaive branch, more correctly called the Cairn Valley Light Railway, as it had been authorized under the Light Railway

Act of 1898, the third car took up the regular working of the branch after the inaugural train, worked by Manson 4–4–0 No. 190, had departed. No. 3 had been made with the engine unit separate from the carriage and coupled to it in the usual manner but such were the complaints of distress caused to passengers due to the acute vibration that it was replaced by a 0–4–4T. Nos. 1 and 2 cars continued on their rostered duties for some years. The last to be withdrawn was No. 1 on the Catrine branch. This occurred when the branch was closed as an economy measure during World War I, on December 31, 1916.

Chapter VIII

THE DRUMMOND TREND WIDENS

In the course of the last decade of the nineteenth century two events occurred which were of prime importance to the North British Railway. The first was the opening of the West Highland Railway in August 1894, the fulfilment of dreams by various promoters from 1846 onwards. The latter culminated in this subsidiary of the N.B.R. which was to prove an interesting and spectacular adjunct to it if not a wholly successful business undertaking when viewed from the commercial angle. The new line ran from Helensburgh on the right bank of the Clyde opposite to Gourock to Fort William 100 miles away across the Moor of Rannoch. Steeply graded and sharply curved, the ruling curvature was 12 chains radius, while the gradients were as steep as 1 in 53 for considerable distances, and a maximum speed of 25 m.p.h. was fixed.

In 1894, at the same time as new locomotives were constructed, some new coaching stock was built specially for service on this line. The coaches were bogie saloons of 22 tons weight and were made up in sets of three to which would be added a 15-ton composite from London, a similar one from Edinburgh and a brake van, the whole totalling 120 tons or so.

To haul these trains the 4-4-0 engines had two inside cylinders 18 × 24 in., 5 ft. 7 in. coupled wheels and 3 ft. 6 in. bogie wheels. The slide valves had $1\frac{5}{16}$ in. lap, $\frac{1}{8}$ in. lead and 4 in. travel in full gear. The exhaust passages jacketed the cylinders and the pistons were conical, perpetuating Drummond's designs.

The boiler barrel was 4 ft. $6\frac{1}{4}$ in. diameter and 10 ft. $2\frac{1}{4}$ in. long with 236 × $1\frac{3}{4}$ in. tubes, giving 1,130·41 sq. ft. The firebox, which had a grate area of 17 sq. ft., had an area of 104·72 sq. ft., the total heating surface amounting to 1,235·13 sq. ft. The working pressure was 150 lb. per sq. in., giving a tractive effort of 14,797 lb.

Laminated springs were fitted beneath the driving axleboxes and spiral springs beneath the trailers and the bogies had spiral spring side control. To ensure adequate braking power two steam brake cylinders were provided, one on either side of the engine between the driving and trailing wheels and both wheels were blocked in front and in rear. Screw and lever reverse were fitted. The weight distribution of the engine was: 15 tons on the bogie, 14 tons 10 cwt. on the drivers and 13 tons 16 cwt. on the trailers, total 43 tons 6 cwt. The tenders carried 6 tons of coal and 2,500 gallons of water and weighed 28 tons 14 cwt. in working order.

In 1893 Nos. 693–8 were built at Cowlairs, followed the next year by Nos. 55, 699–701, 394–5, and in 1896 by twelve more, Nos. 227, 231–2, 341–6 and 702–4. Of these 24 engines 16 were scrapped before the 1923 grouping and 7 passed into L.N.E. stock in their original saturated state, but one, No. 695, was rebuilt in 1919 with a superheater boiler, 10-in. piston valves and entirely new frames to allow for the coupled wheelbase to be lengthened from 8 ft. 2 in. to 9 ft 1 in. The boiler was pitched 10 in. higher than the original and the cylinders were replaced by a new casting having 19 in. bore and 26 in. stroke. The original cab was of the Stirling pattern, but the replacement was of square-cut design with a window on each side. In 1936, under L.N.E. auspices, it was again rebuilt, this time reverting to a saturated boiler with $220 \times 1\frac{7}{8}$ in. tubes and 175 lb. per sq. in. pressure, raising the tractive effort to 20,838 lb. This was not scrapped until 1943.

The second big event in the Holmes era was the race to Aberdeen, in which once again the East Coast and West Coast partnerships were involved, though this time the North British were committed to take the 8 p.m. 'Tourist' from King's Cross on its last lap from Edinburgh to Aberdeen. The company was fully equipped to handle this traffic. Since the opening of the Forth Bridge in 1890 revenue from the new route to Perth and to Aberdeen had been building up and the '633' class of 4-4-0 was coping magnificently. These were designed for the north road following the opening of the Forth Bridge and were slightly larger than the '592' class. Twelve engines built at Cowlairs in 1890 were followed by twelve more in 1,894–5. 18 × 26 in. cylinders, a 20 sq. ft. grate and 1,262 sq. ft. of heating surface were the main improvements on the '592s'. On one run during the race an engine of this class took the 'Tourist' from Waverley to Dundee, $59\frac{1}{4}$ miles, in 59 minutes. The normal schedule was 75 minutes.

The results of the racing were to be felt for long afterwards and almost immediately there was an increase in the loading of trains between England and Scotland by both East and West Coast routes.

Holmes met the call for greater power by bringing out his '729' class of 4-4-0 in 1898, 12 being built in that year (Nos. 729-740) and six more (Nos. 765-770) in 1899. It was also the N.B. answer to the Caledonian challenge as exemplified by McIntosh's 'Dunalastair' of 1896.

These new N.B. engines were almost exact copies of the Caledonian's new ones, but with the boiler pressed to 175 lb. per sq. in.

The boilers contained 254 × 1¾ in. tubes giving 1,224 sq. ft. of surface, only 60 sq. ft. less than the Caledonian engine's 265-tube boilers. Theoretically the N.B. engine would be about 40 tons better than the Caley in haulage capacity. However, in actual fact, though the engines did on many occasions haul trains of 280–320 tons, there is little factual evidence of their work, and in any case by the time Holmes's '729' class made their début McIntosh had got his second 'Dunalastairs' on the road and these were of even greater power.

In 1900-1 Holmes brought out the first six-coupled tank engines since Drummond's engines of 1878. Neilsons built 20, Nos. 795-814, and Sharp, Stewart Nos. 815-834. This class of 40 engines was required for passenger and goods shunting at most big centres and had 17 × 26 in. cylinders, 4 ft. 6 in. wheels and 150 lb. per sq. in. pressure. The boilers were larger than those of the 1878 engines, having 950 sq. ft. of tube surface and 100 sq. ft. of firebox area, while the grate area was 17 sq. ft. At 45 tons 5 cwt. they carried 800 gallons of water and 30 cwt. of coal. When in 1909 the decision to dispense with the Cowlairs engine and rope haulage on the incline was taken, some of the class took up banking duties out of the Glasgow terminus. Later the same year they were replaced by Reid's 0-6-2T designed for this work. In appearance the Holmes engines were not unlike Stroudley's 'E' class on the L.B. & S.C.R. There was in any case a family likeness, since they were based on Drummond's design.

Just as the '729' class was the reply to the Caledonian 'Dunalastair', so the next engines, also 4-4-0s, were the answer to the third 'Dunalastair' class on the neighbouring line. Again there was a similarity of dimensions, 19×26 in. cylinders, 6 ft. 6 in. coupled wheels at 9 ft. 6 in. centres and a total engine wheelbase of 23 ft. 7 in.

The boilers had a total heating surface of 1,577 sq. ft., of which the firebox furnished 133 sq. ft. The grate area was 22·5 sq. ft. The tubes were $1\frac{7}{8}$ in. diameter and 254 provided 1,444 sq. ft. The working pressure of 200 lb. per sq. in. was very high for the period; it had previously been used in Scotland only by Drummond during his experiments on the Caledonian in 1889; in 1906, i.e. three years after production of these engines, the pressure was dropped by 10 lb. per sq. in. The frames of this class were lighter than those of previous designs, as they were made of high tensile steel with a high carbon content. This resulted in a number of frame failures and the consequent withdrawal of the class comparatively early. Some of this trouble may have been due to the stresses initiated on the sinuous Aberdeen road on which they worked. But for Holmes's death in 1903 it is probable that more of them would have been built before the troubles manifested themselves. As it was, only 12 were built and they were numbered 317–328. Nine survived to be taken into L.N.E. stock in 1923.

When deputizing for his chief during the latter's illness Reid introduced some of his own ideas, notably the square cabs and the side windows.

During his superintendency Holmes carried out some rebuilding of older engines. Those of the later E. & G. engines which had not been so treated by Drummond were taken in hand as well as the '382' class 2-4-0 and a number of the double-framed goods engines built at St Margaret's works during Hurst's and Wheatley's eras. One of his actions was to remove the names Drummond had so characteristically bestowed on his engines and none was again named until during the Reid regime.

In 1902 he began the rebuilding of the 'Abbotsfords'. They were given '729' class boilers, but retained the Drummond front-end framing arrangement and tenders. Before the 1923 grouping five of these 12 engines were scrapped, three rebuilt as '729s' passed into L.N.E. hands and the remaining four were rebuilt by Reid also with '729' boilers in 1904 when they were fitted with his side window cabs. The famous *Crampton*, No. 55, was rebuilt for the last time when Holmes gave it a repaired Drummond boiler in 1897 and, at long last, a cab. In this guise it lasted until scrapped in 1907.

Matthew Holmes's health declined so seriously that in May 1903 he resigned from the service and on July 3rd he died in his home 'Netherby' in Lenzie. Deeply mourned by colleagues and work-people alike, his remains were interred in Dalry Road cemetery, Edinburgh.

During Holmes's illness his assistant William Paton Reid had deputized for him and after the former's retirement carried on in the same capacity until his appointment as Holmes's successor was confirmed in June 1904.

Reid was born in Glasgow on February 2, 1862 and he entered the N.B.R. at Cowlairs in May 1879, where his conduct and ability soon brought him to Holmes's notice and he was marked for early promotion. In 1883 he became locomotive foreman at Balloch, in 1889 he moved to a similar position at Dunfermline, thence to Dundee in 1891 and nine years later he was appointed to St Margaret's Edinburgh, the second largest shed on the system. He remained here three years and then became outdoor assistant to Holmes.

It was about this time that big developments were taking place in Scottish locomotive affairs. McIntosh had brought out his huge 4-6-os, Nos. 49 and 50, on the Caledonian, and the G. & S.W.R. had followed with its '381' class 4-6-0. The stage was being set for inter-company competition for the Aberdeen and Perth traffic, additionally there was the need for maintaining N.B. prestige, whose largest locomotives were the '317' class 4-4-os. There was the further factor that train loads were increasing due to the greater use of heavier stock in the form of dining cars and corridor coaches. Reid's first tentative proposal was for a 4-6-0 design comparable with those of the neighbouring companies but his directors rejected the idea, not favouring an engine of such length, particularly in its rigid wheelbase, because of the severe curvature of the 'Waverley' route to Carlisle and the north main line across Fife and the Mearns to Aberdeen.

At the same time, locomotive engineers in Britain were showing a growing interest in the principles of compounding. The De Glehn system in France and the Von Borries system in Germany were finding supporters in Great Britain. Reid examined the compounding arrangements which were being considered by Worsdell on the North Eastern Railway and by Robinson on the Great Central. A design for a compound having one 19 in. high-pressure cylinder and two low-pressure cylinders was prepared and considered alongside a design for a two-cylinder simple engine. The coupled wheels were to have been 6 ft. 9 in. diameter and the working pressure 225 lb. per sq. in. However Reid was not greatly in favour of compounding on the grounds that maintenance costs would be disproportionately high, that friction and back-pressure would both be unduly increased; furthermore compound engines would only be economically sound on runs of over 100 miles, of which the North British had none. Reid was ably supported by Walter Chalmers, his chief draughtsman.

The wheel arrangement of the compound designs of the English companies which Reid consulted was the 'Atlantic' or 4–4–2 type and in view of the board's opposition to the 4–6–0 type the 'Atlantic' would be a necessity if he was to produce a locomotive of the desired power.

The resultant engine was a massive 'Atlantic' type of which fourteen were built by the North British Locomotive Co. in 1906. These engines were true 'Atlantics', i.e. they had outside cylinders and front coupled wheels. The sheer majesty of this design impressed itself on all beholders, laymen and railwaymen alike; moreover in practice they proved excellent machines which were well favoured by the men to whom they were allocated.

The original boilers were 5 ft. 6 in. diameter and 15 ft. long containing 257×2 in. tubes, while the firebox was 9 ft. long and was a narrow box sloping upwards over the trailing truck. The heating surfaces were: tubes 2,071·4 sq. ft., firebox 184·8 sq. ft., total 2,256·2 sq. ft., and the grate area 28·5 sq. ft. The boiler pressure was 200 lb. per sq. in. and three lock-up valves were mounted triangularly on the firebox. A small dome and an even smaller chimney crowned this large boiler which was pitched 8 ft. 11 in. above rail level.

The cylinders were 20 in. diameter and had the long stroke of 28 in. They were inclined at an angle of 1 in 48. The 6 ft. 9 in. coupled wheels gave a tractive effort of 23,506 lb. The weight distribution was 15 tons 18 cwt. on the bogie which had a 6 ft. 6 in.

wheelbase, 40 tons on the coupled wheels spaced at 7 ft. 3 in. centres, and 8 ft. 3 in. behind them was the trailing truck with 18 tons 10 cwt. resting thereon. The side play of the bogie was 2 in. and the trailing truck was also given adequate sideplay. Smith's (North Eastern Railway) piston valves were used and steam reversing gear was fitted. The smokeboxes had no wing plates and the cabs, directly inspired by Worsdell's cabs on the N.E.R., had two windows on each side.

The tenders ran on six wheels and carried 4,240 gallons of water and 6·9 tons of coal, weighing in all 45 tons 8 cwt. which, with the engine total of 74 tons 8 cwt., made a total of 119 tons 16 cwt. The normal load for these engines on the hilly 'Waverley' route was 240 tons.

So successful were these engines that six more were built in 1911, the order being given to Robert Stephenson & Co. These bore the numbers 901–6 and were the same as the originals, which were numbered 868–881, save for the slight modification of certain details.

In 1910, in an endeavour to sway opinion in favour of a 4–6–0, a test run was arranged to compare the performances of one of the 'Atlantics' and an L. & N.W.R. 'Experiment' 4–6–0. On October 23, 1910, each engine, L. & N.W.R. No. 1483, *Redgauntlet*, and N.B.R. No. 881, *Borderer*, ran from Carlisle to Preston and back with a trailing load of 284 tons. This route was chosen because the 'Waverley' route was unsuitable for the English engine on account of inadequate clearances. Maximum speeds ranged between 71 m.p.h. and 80 m.p.h. and drawbar pulls at starting measured 8¾ tons and 8½ tons in the up direction and 11⅛ tons and 10½ tons in the down direction for the 'Experiment' and the 'Atlantic' respectively. In the matter of coal consumption, however, the 'Experiment' with its shallow grate proved far superior to the 'Atlantic' with returns of 58 lb. per mile against 71 lb. per mile. From this and their normal workaday performance the big N.B. engines got a name for being coal-eaters but later, when superheated, the design and designer were completely vindicated.

Having produced his masterpiece for the Anglo-Scottish services between Carlisle and Edinburgh, and Edinburgh and Aberdeen, Reid turned his attention to the growing perishable goods traffic generated in several areas covered by the system, which required engines capable of handling the work both speedily and economically. The first class of engine he produced for this type of work was a 4–4–0 with 6 ft. coupled wheels and known as the 'K' class 'Intermediates', the term here being used to designate mixed-traffic locomotives since they were to be utilized on passenger work as well as on goods, express freight, fish, and perishable trains.

In 1906–7 Nos. 882–893 were built with 19 × 26 in. cylinders and 8¾ in. piston valves. The boilers were pressed to 190 lb. per sq. in. and housed 285 × 1⅞ in. tubes; Reid was using 1⅞ in. tubes in his engines in preference to his predecessor's 1¾ in., and the heating surface was thus 1,620 sq. ft. by tubes, 140 sq. ft. by firebox giving a total of 1,760 sq. ft. and the grate area was 22·5 sq. ft. The engines weighed 53 tons of which 35 tons 14 cwt. was available for adhesion, the tenders at 40 tons carried 3,525 gallons of water and 7 tons of coal.

Rebuilding did not become necessary until 1923–4. In those two years the whole class was fitted with superheaters and the Westinghouse brake fittings were removed, in common with N.B. policy of the period.

These very good engines were followed by a second 'Intermediate' class in 1910–11 very similar in dimensions. The boilers, however, at 5 ft. diameter were 1¼ in. smaller and had 258 × 1⅞ in. tubes—1,478·35 sq. ft. which with 139·78 sq. ft. of firebox surface totalled 1,618·12 sq. ft. and the grate area was 21·13 sq. ft. Cylinders, wheels, piston valves and working pressure were all as before, as were the wheelbase and length. The engine weight was 54 tons 1 cwt. and that of the tender 44 tons 12 cwt. This increase was due to the water capacity being enlarged to 4,235 gallons.

The piston valves used at this time were of the pattern developed by W. M. Smith of the N.E.R. and were of the outside admission type. They were retained after the fitting of superheaters as in the '317', '882' and 'Scott' classes. Lubrication of the fore end was by means of Detroit two-feed hydrostatic lubricators until the introduction of Wakefield four-feed mechanical lubricators in 1919. Like the first 'Intermediates' these engines were dual fitted but the Westinghouse equipment was removed in 1935–6 when steam and vacuum brakes were substituted.

In 1910 No. 867 figured in the comparative trials with the

Highland 'Castle' class locomotive No. 146. It is said that the N.B. engine put up the better performance but no data filtered through to the technical press and official records are also silent regarding the tests.

The manner in which the 1906 'Intermediates' performed their duties prompted a development with slightly larger wheels to handle the passenger work of major importance not covered by the 'Atlantics'. So, in 1909, Cowlairs built Nos. 895–900 and followed them in 1911 by another ten the numbers of which were erratic. Cylinders, valves, motions and boilers were identical with the second 'Intermediates'. The coupled wheels were 6 ft. 6 in. diameter and the tractive effort 19,434 lb. The engines weighed 54 tons 16 cwt. with 36 tons 8 cwt. on the coupled axles, and the tenders, similar in capacity to the 6-ft. engines' tenders, now turned the scales at 46 tons. The class became known as the saturated 'Scott' class since the naming of engines was recommenced and the names were all taken from the novels of Sir Walter Scott.

One of the 1909 group was fitted with an experimental Phoenix superheater in 1911 but the general conversion of the class as a whole was not effected until 1925 onwards. However the 1911 order was for twelve engines but only ten were built at this time. The remaining two formed the first two engines of the superheater 'Scott' class when they emerged from Cowlairs in 1912.

Reid built one class of goods engine in the period 1906–13. They were saturated engines but the fitting of superheaters commenced in 1920 save for an experimental Phoenix type superheater in 1911 on No. 329.

In 1906 Cowlairs built two and in 1908 six. They had Smith piston valves 8¾ in. diameter, and in 1908 six more were built with slide valves, followed in 1910–11 by twelve and ten in 1912–13; the North British Locomotive Co. built ten piston valve engines and ten slide valve engines in their Atlas works in 1906 and 1909, ten slide valve engines at the Queen's Park works in 1909 and ten at Hyde Park works in 1910; a total of 76 engines.

The cylinders of these were 18½ × 26 in. and the wheels were the standard 5 ft. diameter. The boilers were 5 ft. 4¼ in. diameter with 303 × 1⅞ in. tubes, except the two Cowlairs engines which had 338 × 1¾ in. tubes, The fireboxes of these two had a surface

10 sq. ft. less than the standard 143 sq. ft. of the remainder of the class. The grate also varied, 19·25 sq. ft. in the two 1906 engines and 19·8 sq. ft. in the other seventy-four. In all, the pressure was 180 lb. per sq. in. Weights varied between 48 tons 14 cwt. and 50 tons 9 cwt. The tenders carried 3,500 gallons of water and 4½ tons of coal.

The first fifty-four engines were fitted with Westinghouse brakes, the remainder had steam brakes, twenty-seven had vacuum brakes also for working foreign stock, and latterly all the engines of the class were fitted with steam and vacuum brakes. These were Reid's only saturated goods engines.

In 1909 some tank locomotives for suburban and branch line working were constructed in N.B.L. Co's Hyde Park works. At the time they came out these twelve 0-4-4Ts were considered a most modern design. They were put to work on local trains to North Berwick and Musselburgh, to Craigendoran and Arrochar, and other services around Glasgow and Edinburgh. The cylinders were 18 × 26 in. and the coupled wheels 5 ft. 9 in. diameter. The boilers were 4 ft. 8¼ in. diameter and the barrels 10 ft. 2¼ in. long containing 252 × 1¾ in. tubes; these gave 1,214 sq. ft. of surface which, with the 95 sq. ft. of the firebox, made up a total of 1,309 sq. ft. The grate was 16·6 sq. ft. in extent. These boilers were standard with a number of other classes: 4-4-2T, Holmes's 0-6-0 of 1888 as rebuilt by Reid, and the 0-6-2T. The weight was 58 tons 6 cwt. of which 22 tons 10 cwt. rested on the trailing bogie. Coal and water capacities were 56 cwt. and 1,345 gallons.

In appearance these engines very obviously owed much to their Drummond origins, particularly as regards the combined leading splashers and sandboxes, the front faces of which formed the wings of the smokebox front plate. They also had the characteristic rounded edges to the side tanks. Holmes type lock up safety valves were mounted on the domes.

Much the same comments might be made of the 0-6-2T built in large numbers between 1909 and 1924. In 1909 it was determined finally to do away with the Cowlairs hauling rope, and to this end, in order to provide adequate engine power to cope with the severe grade of the incline, the decision to utilize banking engines was made. Six engines with 18 × 26 in. cylinders, 4 ft. 6 in.

30. Glasgow & South Western Railway, rail motor No. 1, 1905.
     I. C. Gillespie's collection.

31.   Glasgow & South Western Railway, Mauson steam rail motor No. 3,
built 1905 for Cairn Valley Light Railway.
      J. F. McEwan.

32. Locomotive Superintendent, Glasgow & South Western Railway,
James Manson 1890–1911, with Mrs. Manson and friend
outside their home in Kilmarnock, *c.* 1906.
Author's collection.

33. North British Railway, 4–4–2T No. 1, Yorkshire Engine Co., 1911. Works No. 1066. *Central Library, Sheffield.*

coupled wheels and boilers similar to those of the 0–4–4T were built by N.B. Locomotive Co. followed by a second six shortly afterwards. So useful and versatile did they prove that a further 94 were built in the ensuing years and employed on goods shunting throughout the system. The bankers were all fitted with Westinghouse brakes but all the remainder had steam brakes only. A significant departure from tradition was that the side tanks were flat-topped instead of having the usual rounded edges. As with the 0–4–4T, new boilers were made and fitted in 1924 with $240 \times 1\frac{3}{4}$ in. tubes instead of the original 252, and in 1938 under the L.N.E.R., a further reduction to 237 tubes was made. At the same time the grate area was reduced to 15·8 sq. ft.

The first engine built in Cowlairs works in 1844 was Paton's bank engine *Hercules*, a six-coupled tank engine, and the last engine to be built in the same works was No. 9227, one of the final order of twenty 0–6–2Ts and the only batch not built by outside contractors, the majority having been built by N.B. Locomotive Co. besides ten from Robert Stephenson & Co. in 1923.

The 0–4–4T engines built in the Hyde Park works in 1909 were not the complete answer to the traffic department's needs for suburban and other local services so that, in 1911, a new departure (for Scotland at least) made its appearance. The Drummond 4–4–0T of 1880 was the direct ancestor of the new engines which were 4–4–2Ts. They were more than just Drummonds with the addition of a trailing truck, 18×26 in. cylinders, Stephenson link motion and slide valves as in so many N.B.R. types, together with 5 ft. 9 in. coupled wheels and 175 lb. per sq. in. pressure gave an engine with the same power as the 0–4–4T. The water capacity of 1,990 gallons ensured greater range of activity as compared with the 1,345 gallons of the 0–4–4T and coal space for 4 tons instead of 56 cwt. Good flexibility was arranged for by allowing the bogie 1 in. side play and the trailing truck 3 in. The latter was in the form of a radial axlebox which was pivoted 7 ft. 6 in. in advance. These engines were built by Yorkshire Engine Co., in their Meadowhall works at Sheffield, the first eleven in 1911–12, followed by nine in 1912–13 and a further ten in the latter year. They were all dual fitted but whereas the first eleven had the Westinghouse pump fitted on the front of the right-hand side tank, the remainder had it attached to the side of the smokebox. Carriage warming apparatus was fitted with pipes front and rear and a Wakefield hydrostatic lubricator was fitted in the cab.

The man who followed David Jones at Inverness in 1896 was Peter Drummond, younger brother of Dugald. Peter was born in 1850 and served his apprenticeship with the Glasgow engineering firm of Forest & Moor, on completion of which he went to Brighton where he joined his brother in the service of the L.B. & S.C.R. When Dugald was appointed to the N.B.R. in 1875 Peter also went to Cowlairs, and in 1882 both brothers crossed the lines to the Caledonian works where the post of under-manager was specially created for the younger man. This was not included in the establishment but it was a post in which he was able to gain a large amount of experience under the guidance of the redoubtable Dugald and his chief draughtsman Tom Weir.

Hitherto the Highland had been an outside-cylinder engine line but with the coming of Peter Drummond to Lochgorm the pattern changed. With his first engines he introduced inside cylinders to the Highland Railway. These were so similar to the new engines his brother had just put into service on the L. & S.W.R. that they might be considered to have been made from the same drawings. The cylinders were however increased in diameter to $18\frac{1}{4}$ in. and the stroke to 26 in. The only other Highland engines using a 26 in. stroke at that time were the 'Big Goods' after which Jones had reverted to 24 in. stroke. It is noticeable that the three companies on which the Allan influence had persisted were the last to adopt the longer stroke and therefore higher piston speeds, having been content with moderate piston speeds and possibly less difficulty with cylinder lubrication. The new Highland engines had 6 ft. coupled wheels and 3 ft. 6 in. bogie wheels. As in so many Scottish designs the bogie centre pin was placed 1 in. forward of the centre line between the axles. The tube heating surface was derived from $214 \times 1\frac{3}{4}$ in. tubes, 1,060 sq. ft., the firebox gave 115 sq. ft., so that the total was 1,175 sq. ft. with a 20·3 sq. ft. grate.

A departure from previous Highland practice was the use of Stephenson link motion instead of Allan's straight link gear and the double-beat regulator valve was supplanted by a slide valve type. The safety valves were of the lock-up type and were mounted on the dome, set to blow off at 175 lb. per sq. in.

The first engine delivered was No. 1, *Ben Nevis*. Although photographed in this guise it never ran thus, being immediately renamed *Ben-y-gloe* by order of the board who objected that Ben Nevis was outside Highland territory. Nos. 2 to 8 followed, all eight being built by Dübs & Co. in 1898–9. Lochgorm works built Nos. 9–11 in 1899 and six more in 1900–1. Three further engines were built by N.B. Locomotive Co. Queen's Park works in 1906. The first seventeen engines had 3,000-gallon tenders on six wheels and carried 5 tons of coal. Four of these tenders were exchanged for bogie tenders attached to some of the 0–6–0 goods engines built in 1900; the last three engines built had 3,185 gallon tenders, also on six wheels.

The engines weighed 40 tons 4 cwt. with 30 tons 15 cwt. on the coupled wheels and the two sizes of six-wheel tenders weighed 37 tons 10 cwt. and 38 tons 7 cwt. respectively. Six of the class were fitted with Westinghouse brakes and two were later fitted with F. G. Smith's feed water heating apparatus. The class became known as the 'Wee Bens' to distinguish them from the later class which had larger boilers.

The 0–6–0 goods engines referred to appeared in 1900, also built by Dübs & Co. The first six were Nos. 134–9 and had 40-ton bogie tenders which were of the inside frame variety typical of the ones popular with Dugald Drummond. The cylinders, boilers and motions were identical with the passenger engines. Four more of the goods engines came out in 1902 and another two in 1907.

The 1902 engines had cross water tubes in the firebox for the purpose of improving the circulation of the water and the transference of heat. These cross water tubes were retained by Drummond's successors but were removed by the L.M.S. authorities when rebuilding was carried out in 1923–34.

All the engines of this class were fitted with vacuum brakes only. Of the thirty engines of both classes sixteen of the passenger and ten of the goods were reboilered with boilers of the Caledonian type in the period 1926–37 and one of the goods engines was so dealt with as late as 1945. Scrapping started in 1933 but nine 'Bens' and seven 'Barneys', as the goods engines were called, lasted until after nationalization in 1948. When this reboilering was being carried out the safety valves were taken off the domes, repositioned on the firebox, and substituted by Ross Pop valves. At the same time the wings of the smokebox front plate were removed.

At the time of the accident to his leg, which resulted in the premature resignation of David Jones the success of the 'Big Goods' was undoubted. They had been in service a bare two years but had proved equally good machines on either goods or passenger work, though originally intended for the former. On the strength of this Jones proceeded to design a passenger 4–6–0 on the same general lines as the goods engine and this design was well advanced when his mishap occurred. It was therefore, left to Drummond to complete the preliminaries and get the order for the construction of the new engines placed with the manufacturers. The contract was given to Dübs & Co. who turned out the first six, Nos. 140–5, in 1900.

Many Jones features were retained. The outside cylinders $19\frac{1}{2} \times 26$ in., Allan straight link motion, and the boiler, were all based on previous practice but Drummond introduced his own marine-type big ends, cab and boiler mountings, Richardson balanced slide valves and Dugald's steam reverser. He also perpetuated the practice of putting the bogie centre pin 1 in. in front of the centre line.

The boiler was 4 ft. $9\frac{3}{4}$ in. diameter and 14 ft. $4\frac{1}{2}$ in. long and contained 248×2 in. tubes equal to 1,916 sq. ft. of surface, while the firebox contributed 134 sq. ft. in a total 2,050 sq. ft. The grate area was 26·5 sq. ft. and the working pressure 175 lb. per sq. in. The tractive effort was thus the respectable figure of 2,1325 lb. The engine weight was 59 tons 18 cwt. 2 qr. with 15 tons 1 cwt. on the bogie, 15 tons 1 cwt. on the leaders, 15 tons 3 cwt. on the drivers and 14 tons 13 cwt. 2 qr. on the trailers. The tender had a coal capacity of 5 tons and of water 3,350 gallons, It was carried on two bogies and was very like the L. & S.W.R. bogies in appearance.

These engines were put on the Inverness–Perth route via Carr Bridge and were an immediate success. The class was increased by four in 1902, two in 1910–11 and a further four in 1913.

During World War I, when the Highland Railway motive power matters were at a very low ebb, three generally similar engines with certain modifications were built at the N.B. Locomotive Co. Queen's Park works, the one-time Dübs establishment.

The excellent performance of the original engines led to the purchase of Fifty similar machines by the C. de F. l'Est of France. They had Westinghouse instead of vacuum brakes and were built by N.B. Locomotive Co. Under L.M.S. management three of these engines, which were named after castles in the Highlands, and therefore gave the class the name 'Castle' class, were given new boilers in which there were 292 × 1¾ in. tubes, 1,974, sq. ft., and the fireboxes were fitted with direct roof stays in place of the original girder stays. This work was done at St Rollox and the new boilers were of the Caledonian type but with fittings from other sources, e.g. Derby type water gauges and frames and Horwich type pop valves. When transferred to the Oban line these three engines gave a good account of themselves. Two of the 'Castles' were singled out for special treatment; one was experimentally fitted with a Phoenix superheater for a while and another, No. 146, was selected for some comparative trials against a Reid 'Large Intermediate' from the N.B.R. The sections of road over which the tests were made were from Blair Atholl to Dalwhinnie on the Highland main line and from Perth to Kinross on the N.B.R.

In 1903 two 0–6–0 side tank engines appeared from Lochgorm works, a third following in 1904. These were Nos. 22–4. They were built up from the cylinders and motions of old 2–4–0 goods engines which were being broken up, and the boilers of 2–4–0 passenger engines. The tank engines, as reconstructed, had 18 × 24 in. cylinders, 5 ft. 3½ in. wheels and 1,186 sq. ft. heating surface of which the firebox contributed 93 sq. ft. The grate area was 16·2 sq. ft., the pressure 160 lb. per sq. in. and the weight 47 tons 18 cwt. The tanks held 1,000 gallons and the bunkers 35 cwt. of coal. These three shunting engines, which were steam brake fitted, lasted into L.M.S. days and were scrapped in 1930–2.

Branch lines were catered for by three very neat 0–4–4Ts which came from Lochgorm works in 1905. A fourth was added in the following year. These somewhat diminutive engines, with cylinders 14 × 20 in., coupled wheels 4 ft. 6 in. diameter and 2 ft. 6 in. bogie wheels, had a total wheelbase only 18 ft. 3 in., made up of 6 ft.

between the coupled axles and 5 ft. between the bogie axles. The boiler had 170 × 1¾ in. tubes, 652 sq. ft. and the firebox gave 67·5 sq. ft. with a 13 sq. ft. grate. The working pressure was 150 lb. per sq. in. and the coal and water capacities 25 cwt. and 900 gallons whilst the weight in working order was no more than 37 tons 15 cwt.

Drummond's final tank design was a large 0–6–4 intended for banking duties on the 'Hill' but was also used on local passenger and goods workings. The N.B. Loco. Co. built eight in their Queen's Park works: four, Nos. 39, 64, 65 and 69, in 1909 and four, Nos. 29, 31, 42 and 44, in 1911. The dimensions of these 69½-ton engines were in keeping with what was expected of them. The 18¼-in. cylinders with 26 in. stroke and 5 ft. coupled wheels gave a tractive effort of 22,077 lb. at 85 per cent of the 180 lb. per sq. in. pressure. The boilers were smaller only than the 'Castles' and the last 4–4–0s, the 'Big Bens' which appeared about the same time; the total heating surface of 1,268 sq. ft. was made up of: 230 × 1¾ in. tubes 1,148 sq. ft., firebox 120 sq. ft., while the grate area was 22·43 sq. ft. The tanks held 1,970 gallons of water and the bunkers 4½ tons of coal. In 1932 scrapping started and by the end of 1936 all had gone.

In 1908 Queen's Park turned out Drummond's last Highland engines, the 'Big Ben' 4–4–0 class. It was based on the 1898 'Wee Bens' and betrayed a family likeness to the '395' class Dugald had introduced on the L. & s.w.R. Cylinders, motions and frames were similar to the original 'Bens' but the boilers differed in several respects. The diameter was increased to 5 ft. 1¾ in. and the number of tubes to 266. As these were 2-in. diameter, the tube surface was 1,516·2 sq. ft. The firebox surface was 132 sq. ft. in a total of 1,648·2 sq. ft. The grate area remained 20·3 sq. ft. but the working pressure was raised to 180 lb. per sq. in. These engines weighed 52 tons 6 cwt. of which 17 tons 2 cwt. was on the bogie, 17 tons 16 cwt. on the driving axle and 17 tons 8 cwt. on the trailing axle. The first four engines were turned out with 3,185-gallon six-wheel tenders, but Nos. 61 and 63 were later given the 3,200-gallon bogie tenders taken from two 'Barneys'. Nos. 66 and 68 retained their six-wheel tenders but the last two built, Nos. 60 and 62, had even larger bogie tenders of 3,600 gallons capacity weighing 51 tons 14 cwt. For working foreign stock Nos. 60, 61 and 63 were fitted with Westinghouse brakes, and in the Smith regime the class was fitted with that engineer's feed water heating apparatus. All six engines were built at the N.B. Locomotive Co's. Queen's Park works, the first four in 1908 and the last two in 1909. Just prior to the 1923 Grouping when D. C. Urie was the Highland's last locomotive superintendent, drawings were prepared for a superheater boiler and the alterations were made between 1924 and 1927. The class was scrapped from 1932–37.

After delivery of the second batch of 0–6–4Ts in 1911 no further engines appeared for over twelve months and in February 1912 Peter Drummond left Inverness to take up a similar post on the G. & S.W.R. vacant on the retirement of James Manson.

When Dugald Drummond left the Caledonian Railway he was followed at St Rollox by the short-lived Smellie who was in his turn succeeded by another Ayrshire man, John Lambie. Lambie had been born in Saltcoats in 1833 and was thus seven years older than Smellie and in his fifty-eighth year when he took office.

Whilst he was still a child Lambie's father had taken the family from Saltcoats to Motherwell where he had been appointed traffic manager of the newly established Wishaw & Coltness Railway. When he was twelve years old the young Lambie started as an apprentice in the Holytown Bridge shops of the W. & C.R. He continued in the service of this company which was first leased, then bought outright, by the Caledonian and he rose in the department until he was chosen by Drummond to be his assistant superintendent. His reign as locomotive chief of the Caledonian was however destined to be a short one, for he too died comparatively early, in February 1895. Yet in the space of three years and ten months he contributed five designs to the Caledonian.

Smellie came from a line where the use of $1\frac{5}{8}$ in. tubes was standard practice. The Caledonian standard was $1\frac{3}{4}$ in., yet the first two Lambie designs, a 4–4–0T and a 0–4–4T, both employed $1\frac{5}{8}$ in. tubes. Both engines were required for the Glasgow Central Railway, the East–West suburban line which ran mainly beneath the city streets on the cut-and-cover plan. The line was authorized in 1888 and was opened in sections from 1894–6.

The 4–4–0Ts were built at St Rollox in 1893 and were numbered 1–12. 17×24 in. cylinders, 5 ft. coupled wheels and a working pressure of 150 lb. per sq. in. produced a tractive effort of 14,730 lb. The heating surface, 1,095·76 sq. ft., was made up of tubes 894·6 sq. ft. and firebox 111·1 sq. ft. The grate area was 17 sq. ft. and the coal and water capacities 2½ tons and 1,000 gallons. They weighed 50 tons 6 cwt. 3 qr., of which 15 tons 9 cwt. was on the bogie.

In 1895 the 0–4–4Ts, Nos. 19–28, were turned out of St Rollox works. The boilers were identical with the 1893 engines but the cylinders were 18×26 in. and the coupled wheels 5 ft. 9 in. diameter. These engines were three tons heavier than the earlier ones and carried the same amount of water but only 2 tons of coal. Both classes were fitted with condensing apparatus but even so the Central Underground stations were unhealthy places and redolent of smoke and all the other smells concomitant with an underground railway.

Goods traffic on this line, mainly transfer traffic from goods depots, foundries, etc., on the East side of the city to depots and shipyards on the West side, was handled by an 0–6–0T class of which nine were built in 1895 to Lambie's designs. Again the cylinders were 18×24 in. but the wheels were no more than 4 ft. 6 in. diameter. The boilers were the only Lambie boilers to have 1¾ in. tubes, quite possibly the result of McIntosh's modification after he had succeeded Lambie. 206×1¾ in. tubes gave 975 sq. ft. of surface, not greatly different from the previous designs, and the fireboxes and grates were the same as in the earlier engines. These engines also had condensing apparatus fitted and a feature of the gear was the specially designed pumps, with a low internal water velocity and an automatic air feed connected to an air vessel to reduce shocks due to high-speed running. The condensers gave rise to quite an amount of trouble. The first ill effect was the heating of the water in the side tanks to such a degree that the injectors failed to work. The steps taken to remedy this were, first, the fitting of pumps worked off the crossheads and Gifford injectors as modified by Drummond; these fittings required a clackbox with double clacks on each side of the boiler. Second, the crosshead pumps were themselves a source of trouble which was remedied by fitting duplex pumps similar to the Westinghouse pumps. Better results accrued but the steaming of the engines suffered, when the condensers were working, due to the depletion of the draught on the fire caused by the diversion of the exhaust steam to the condenser pipes. In consequence, and despite repeated action by higher authority against offenders, the operation of the apparatus was more observed in the breach than in accordance with the instructions.

The same year, 1895, also saw five 0–6–0 saddle tanks built at St Rollox with 18×26 in. cylinders and 4 ft. 6 in. wheels. The boilers were similar to those of the two passenger types.

Lambie's undoubted success, however, was the 4–4–0s which he brought out in 1894. In these he took the basic Drummond engine and only made a few minor changes. The boiler tubes followed the dictates of his predecessor, there being $238 \times 1\frac{5}{8}$ in. tubes giving 1,071·5 sq. ft. in a total 1,184·12 sq. ft. In other respects the dimensions were the same as the Drummonds. Six were built in the company's shops, Nos. 13–18, and it was No. 17 in the hands of driver John Soutar and fireman D. Fenton that made the fantastic running between Perth and Aberdeen in the 1895 race to the latter city.

The Drummond engines which also shone in the race, Nos. 78 and 90, were of the later breed of the standard 4–4–0, and the Lambie development of the type, having the same arrangement of exhaust steam jacketing of the cylinders and divided slide valves ensuring short straight steam passages and free exhaust to the vortex blast pipe, had the advantage over contemporary designs of being extremely free running and economical engines.

The results of the tests carried out by Drummond in 1889 are in the archives of the Institution of Civil Engineers. The best figures obtained were a coal consumption of 42·27 lb. per mile by No. 79 carrying a pressure of 200 lb. per sq. in. The evaporation rate 6·576 of water per lb. of coal was low but explained by the fact that the tests were carried out at a time when stack coal was being used during a colliery strike, and according to Drummond, 'Fifty per cent of it would have passed through a $\frac{3}{8}$ in. mesh riddle. 25 per cent passed into the atmosphere without evaporating a lb. of water.' Despite this handicap this engine developed 806 IHP at 33 m.p.h. when ascending Beattock bank with 140 trailing tons. The superiority of the higher working pressures was clearly demonstrated in comparison with No. 78 pressed to the standard 150 lb. per sq. in., which on the same climb produced 623 IHP at 25 m.p.h. with a 170-ton train.

Lambie ran an interesting test between Glasgow and Carlisle and back with one of his own 4–4–0s of which the number has not been left on record. The trains were the up and down 'Corridor' and the trailing loads 235 and 212 tons respectively. The round trip was performed on a coal consumption of 35·021 lb. per mile and an evaporation rate of 7·51 lb. of water per lb. of coal. These 1894 engines showed a distinct advance on their forerunners particularly when it is remembered that no assistance was taken either up to Craigenhill summit on the up trip, or to Beattock on the down journey, in each case with loads in excess of 200 tons.

Chapter IX

JOHN FARQUHARSON McINTOSH

During the latter part of the nineteenth century a railway enthusiast who was also a Scottish legal luminary, Norman Doran Macdonald, had been bombarding the locomotive engineers of the day for 'Bigger Boilers, Better Brakes'. Some good appears to have resulted from this campaign, or at least to have occurred contemporaneously. The year 1889 saw the compulsory fitting of continuous automatic brakes decreed by government through the Board of Trade, and the last decade of the century ushered in the era of the big boiler.

The man responsible for this development was John Farquharson McIntosh who succeeded to the vacancy created by the untimely death of John Lambie.

Like his predecessor, McIntosh was a 'running man' who had entered the service of the Scottish North Eastern Railway at Arbroath in 1856 when he was fourteen years old. In 1865 he passed out as a fireman and two years later he qualified as a driver and moved to Montrose. Here, in the course of time, he gained good experience with various classes of railway rolling stock and in 1876 was made inspector covering all lines north of Perth, all of which had become Caledonian territory by the absorption of the S.N.E. in 1866. An accident in 1877 resulted in the loss of his right hand and in the same year he enlarged his inspectorate by being given the lines from Greenhill to cover as well as the northern lines. Further promotion saw McIntosh as locomotive foreman at Aberdeen, Carstairs and Glasgow (Polmadie) and later he was chief inspector under Drummond. He continued in this capacity under Smellie and Lambie, whose locomotive running superintendent he became, and from February 1, 1895, he took up the post of chief locomotive engineer of the Caledonian Railway. From the wealth of experience McIntosh had gained in what later generations would term the 'Motive Power Department', save that in his day locomotive running was the province of the locomotive superintendent who took the workshops in his charge as well, he knew precisely what the man on the footplate wanted in order to cope with the ever-growing and heavier traffic.

McIntosh had only been in office six months when the great competitive racing from London to Aberdeen took place, the West coast partners, L. & N.W.R. and Caledonian, producing some pyrotechnics in the form of fast running. And in that spectacular demonstration of just what could be achieved by the engines of the day the Caledonian showed itself to be the complete master of the situation.

The machines which whisked the race trains through the Scottish countryside at speeds averaging more than 60 m.p.h. and often exceeding maxima of 75 m.p.h. were the excellent products of Drummond and Lambie. When it is realized that Archibald Crooks brought a 207-ton train into Perth one morning with a tank that had been dry for the last fourteen miles the clear need for bigger and better engines with tenders of greater capacity becomes apparent.

It has been said that the general trend of the times was to increase the cylinder diameter by half-an-inch and decrease the heating surface by 50 sq. ft. which, of course, rapidly produced conditions found on so many lines, engines without adequate steaming capacity for the size of the cylinders. Now, with the advent of the new superintendent on the Caledonian a new era was to dawn.

McIntosh's first locomotive had a boiler of adequate proportions and these were obtained by pitching the boiler 7 ft. 9 in. above rail level, six inches higher than in the Lambies', thus obtaining a diameter of 4 ft. 8¾ in. over the standard 6 ft. 6 in. coupled wheels. The barrel length remained 10 ft. 7 in. but it contained 265 tubes of 1¾ in. diameter giving a heating surface of 1284·45 sq. ft., an increase over Lambie's of 19 per cent. The firebox had an area of 118·78 sq. ft. and the total was 1403·23 sq. ft. while the grate area was increased to 20·63 sq. ft. The ratio of free gas area to grate area showed an improvement over both Drummonds and Lambies, being 16·1 per cent as compared with 12·9 per cent and 14·4 per cent of the previous engines, so that free steaming was assured. To utilize the steam to best advantage the cylinder diameter was raised to $18\frac{1}{4}$ in., the stroke remaining 26 in. As in previous designs the cylinders were steam jacketed (exhaust steam) and Vortex blast pipes were retained, but the valves were centrally placed, not being separately disposed at each end of the cylinder as had been done by Drummond. The port areas were generous, the steam ports being 9·9 per cent of the piston area and the exhaust ports 19·8 per cent.

McIntosh did not continue the Drummond practice of securing the slide bars at their centres, instead he reverted to the more orthodox method of bolting them to lugs on the cylinder covers and the spectacle plates. Shortly after the engine went into service, fourteen more were built at St Rollox between the end of January and the middle of May 1896; the late Charles Rous-Marten recorded that they were working the West Coast services in the style and at the speed of the 1895 race trains but with *double the loads*.

From the start their success was assured. The first engine, No. 721, was named *Dunalastair* after the Perthshire estate of the chairman, J. C. Bunten.

It might be thought that with this design there was a splendid opportunity for a determined effort at standardization of component parts, but it seems that McIntosh was not yet satisfied that he had reached the state of perfection that would justify such a course. No more of these locomotives were built but an improved larger edition which became known as the 'Dunalastair II' class was turned out of the company's shops in 1897 when Nos. 766–771 appeared, followed by nine more, Nos. 772–780, in 1898.

To obtain greater heating surface the boiler was lengthened by $9\frac{1}{2}$ in. The number and size of the tubes remained the same but the surface was 1381·22 sq. ft. The firebox and grate were as before and the total heating surface, now 1500 sq. ft., was almost 7 per cent more than in the 'Dunalastair I' class. The cylinders were enlarged to 19 in. diameter and the working pressure raised to 175 lb. per sq. in. so that the tractive effort became 17,850 lb. against the 'Dunalastair Is' 15,096 lb. The longer boiler necessitated the lengthening of the wheelbase to 23 ft. 1 in. The additional 12 in. was placed between the bogie centre and the leading coupled axle and had the effect of increasing the length of the connecting rods by 7 in., so reducing the effect of angularity of the rods.

The leading sandboxes of the first Dunalastairs were integral with the driving splashers, Drummond fashion, but in the second generation they were made separate and placed below the running plates allowing the driving and trailing splashers a neat sweeping curve to the front of the cab.

Perhaps the most noticeable feature of the new class was the tender. This was carried on two bogies and was the first use in Scotland of the type, predating Manson's on the G. & S.W. by six years and Peter Drummond's on the Highland by three years. True, Manson had used a bogie on some of his G.N.S. and G. & S.W. 4-4-0s but had fitted one bogie in conjunction with two fixed axles.

The frames of the new Caledonian bogies were outside the wheels and the springing was a unique arrangement incorporating a 17-plate spring slung between the axleboxes in an inverted position so that the buckle transmitted an upward thrust on the centre of an equalizing beam bearing on the axlebox tops.

A test run with No. 772 of this class hauling the up 2 p.m. 'Corridor' express from Glasgow to Carlisle on February 23, 1898, when the engine was three months old, showed that a 305 ton train could be handled with ease. The coal consumption was 49 lb. per mile, or 110 lb. per sq. ft. grate per hour; at this rate the evaporations were 7 lb. water per lb. coal. During the run an indicated horse power of 1,020 was attained at a speed of 50 m.p.h. at the top of the 1 in 348 4½ miles out approaching Cambuslang. On the climb up to Craigenhill with the speed varying between 26½ m.p.h. and 30½ m.p.h. the IHP calculated from the indicator diagrams taken at one minute intervals showed the engine to be developing between 750 and 810 IHP. The cut-off during the climb was kept at 38 per cent until a point nearing the summit, when it was shortened to 31 per cent and held at that during the remainder of the trip. From Carstairs to Beattock summit the output was between 800 and 900 IHP at an average speed of 48 m.p.h. Details of this test were published in *The Engineer* December 23, 1898.

The development of the Caledonian 4-4-0 did not end at this point, however, for in 1899 a still larger version of the 'Dunalastair' issued from St Rollox and became known as the 'Dunalastair III' class. The first three were built in 1899 and were followed by thirteen more the next year.

The machinery of these engines was much the same as before but the boilers were substantially larger. The tube surface was increased to 1,402 sq. ft. by the inclusion of four additional tubes, bringing the number to 269. Whereas the 'Lambies' and 'Dunalastair I' classes used brass tubes McIntosh changed to the use of copper in the second and subsequent developments of his 4-4-0s. In the third series the fireboxes were increased to 138 sq. ft. and the grates to 22 sq. ft. so that the total heating surface became 1,540 sq. ft. The working pressure was raised to 180 lb. per sq. in. and gave a tractive effort to 18,411 lb. The coupled wheelbase and total engine wheelbase were each lengthened by 6 in. to 9 ft. 6 in. and 23 ft. 7 in. respectively, and although the boiler barrel dimensions were the same as in the 'Dunalastair IIs' the centre line was raised a further 3 in. to 8 ft. above rail level.

Bogie tenders were attached to these engines but were 3 in. shorter wheelbase, the difference being in the distance between the bogie centres. Scaling a total of 101 tons 3 cwt. these were the heaviest engines yet built by the Caledonian Railway.

In 1898 McIntosh was approached by the Belgian Government for the supply of locomotives of the '766' class. The directors agreed to the gift of drawings and full particulars of these engines. Neilson Reid & Co. built five under the overall supervision of R. W. Urie, chief draughtsman, who represented McIntosh. In most respects they were identical to the Caledonian prototypes but had to be converted to right-hand drive to comply with the Belgian requirements. Later some 225 additional locomotives were constructed in Belgium to a slightly enlarged design and the McIntosh cult was further expanded in that country in some 0-6-0 goods and 4-4-2 passenger tank engines.

In 1914 the success of superheating had become manifest and a number of the second and third generations of the 'Dunalastairs' were rebuilt with superheaters and those not so treated were given new saturated boilers of the original design but with Ross Pop safety valves and without the wing plates to the smokeboxes so characteristic of Scottish locomotives of the Drummond design trend.

The last of McIntosh's 'Dunalastair' classes was the fourth

series which first came out in 1904. The first two were Nos. 140 and 141 built in May. Three, Nos. 142–4, appeared in June, Nos. 145–150 were built in 1905–6, Nos. 923–7 in 1907–8 and finally Nos. 136–8 in 1910.

Once again the cylinders and motion were to some extent standardized but the boilers were once again enlarged. The barrel diameter was increased to 4 ft. $11\frac{1}{2}$ in. and its length to 11 ft. 2 in. The tubing was altered and two sizes of tube were used: $255 \times 1\frac{3}{4}$ in. and 21×2 in. The tube heating surface became 1,470 sq. ft. which, with the 145 sq. ft. of the firebox produced the high total of 1,615 sq. ft., 15 per cent greater than the original engines of 1896. The grate area was reduced to 21 sq. ft. but this difference was not sufficient to affect the steaming of the boiler. The method of firing which prevailed was to run with a fairly thick fire on a smallish grate, a practice which suited the soft Scottish coals.

The total weight went up to 56 tons 10 cwt. of which the driving and trailing axles carried 37 tons 15 cwt. Similarly the tender weight rose to 50 tons 15 cwt. due to the water capacity being increased to 4,300 gallons. These alterations made the total weight of engine and tender 107 tons 5 cwt. which was a substantial figure for a 4–4–0 in the early years of the present century. There were, as might be expected, a number of features common to all four of these fine 4–4–0s.

The tyre fastenings used by McIntosh were those introduced by Drummond on the N.B.R. in 1875 and which he continued to use at St Rollox in 1882.

McIntosh suspended his brake blocks in front of the driving and trailing wheels with the object of pulling the blocks in one direction only during an application and so reducing the stresses in the coupling rods. Elsewhere, as on the North British Railway for example, the blocks were placed in front of the drivers and behind the trailers.

A further means of reducing stresses in the rods was the practice of fitting the coupling rod crank pins in the wheel centres to give a 10-in. throw.

Ample bearing surface was provided on the axle journals thereby contributing to the freedom from heated axle bearings enjoyed by the whole of the genus 'Dunalastair'. The following table shows how the driving and trailing bearings grew in size with each

34. Highland Railway, 4–6–0 No. 70, F. G. Smith, built Hawthorn Leslie & Co., 1915. 'River Class'—sold to Caledonian Railway. *Author's collection.*

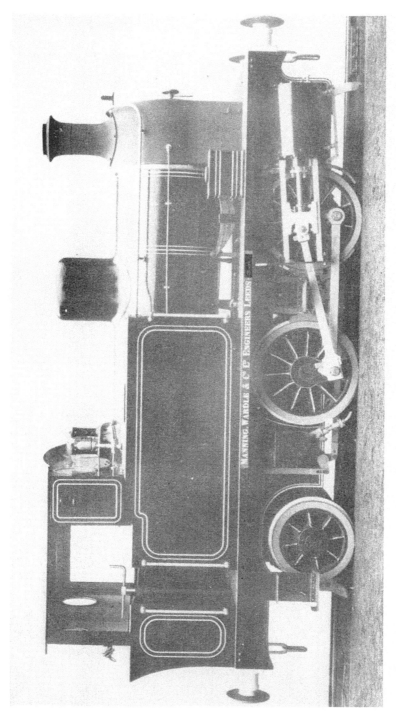

35. Great North of Scotland Railway, Manning Wardle & Co. 0–4–2T, Nos. 1884, 1885, built 1915. *By courtesy of Hunslet Engine Co.*

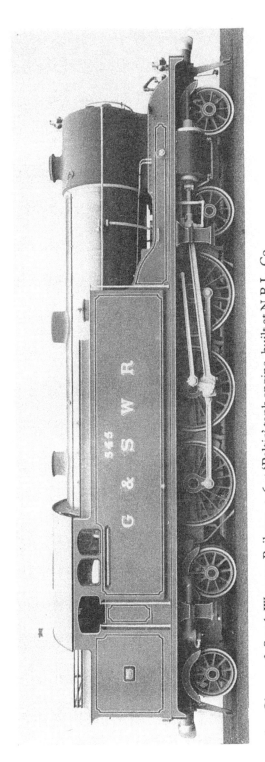

36. Glasgow & South Western Railway, 4–6–4 'Baltic' tank engine, built at N.B.L. Co. Hyde Park Works, 1922, to designs of R. H. Whitelegg. *Mitchell Library, Glasgow.*

37.   Glasgow & South Western Railway, Manson 4 Cyl. No. 11 as rebuilt by Whitelegg, 1922. *D. L. Smith.*

successive development, together with the projected areas. The Pickersgill 4–4–0 of 1916 has been included for the purpose of comparison.

Dimensions of Driving and Trailing Bearings
McIntosh and Pickersgill 4–4–0

Class	Driving in.	Trailing in.	Projected Area Driving sq. in.	Projected Area Trailing sq. in.
Dunalastair I	8 × 7½	7½ × 7½	60	56·25
Dunalastair II & III	8½ × 7½	7¾ × 9	63·75	69·75
Dunalastair IV	9¼ × 7½	9¼–7¾* × 12	69·375	102·58
Pickersgill	9½ × 9	8¼ × 12	85·5	99

* Waisted, 24·375 in. radius

All McIntosh's 4–4–0s followed the Scottish practice of placing the bogie centre pin 1 in. forward of the centre line of the bogie wheelbase but this peculiarity of design was not carried on by Pickersgill when he took over in 1914.

During the protracted colliery strike in 1912 No. 724, one of the 'Dunalastair I' engines, was fitted with the Holden apparatus for burning oil as a fuel as developed by that engineer on the Great Eastern Railway. A fuel tank of 520 gallons capacity was mounted in the tender coal space A trial of a Weir feed water heater and pump was made in 1915; the engine selected for this test was No. 136, one of the 1910 'Dunalastair IV' class.

The springing of McIntosh's 4–4–0 engines was the Arrangement prevailing on other lines at the period. Coil springs under the driving axleboxes and laminated springs under the trailers were used on the Dunalastair I, II and III classes whilst the IVs had laminated springs under all four axleboxes. Unlike the usual practice, the coil springs were two inches greater in diameter at the top than at the bottom. McIntosh patented his own type of boiler water gauge glass protector and used it throughout his designs. The protector was made of toughened glass and had been tested to a pressure of 3,000 lb. per sq. in. without breaking. It was a little more than half a circle in shape enclosing the gauge glass and had rubber washers to prevent damage by jarring.

Double heading of trains was a practice not condoned by McIntosh and, in order to discourage such a practice, the leading drawhook was fitted with a 'D' link only, instead of the usual screw coupling, and the 'D' link was usually to be seen resting against the front platform rather than hanging from the hook.

The intermediate drawbar had the customary pin through the bushes in the engine dragbox; behind the tender dragbox there was a spiral spring secured by a plate and hexagon nut. Spiral springs were also used for the intermediate side buffers and in addition there was a centre buffer on the tender headstock mating with a rubbing plate on the engine.

The staying of the firebox crown plates varied. In the first series girder stays were suspended from two angle-irons. The same arrangement was applied in the second series but the third and fourth series had girder stays slung from two pairs of angle-irons, the front and rear pairs having long pins through them to suspend the girders.

In all four of the 'Dunalastair' classes the regulator was of the double beat type and was housed in the dome.

A sight feed lubricator supplied oil to the pistons and valves and auxiliary oil boxes with worsted tail trimmings provided lubrication for the axle journals.

Steam and lever reversers ensured that the engine could be moved at any time instead of having to await the required amount of steam as in the case of Stirling's engines.

The need for more locomotives to cope with the still growing goods and mineral traffic was met, for a time, by continuing to turn out o–6–os of Drummond's design. These required several modifications to bring them within McIntosh's concepts of design and filled the breach until his own new goods engines appeared in 1899. In all, 83 Drummonds came out between 1895 and 1899; they were dimensionally similar to the originals with some modification of details. Longitudinal girder stays replaced the Drummond direct stays for the firebox crownplate. The motion, or spectacle, plate was brought back from the centre of the slide bars to the rear ends of the bars. A sliding regulator valve gave place to the double-beat type of valve and Furness lubricators were substituted for tallow cups on the smokebox front.

Externally the most noticeable features were the brake pull rods and the tender springs. On the Drummonds the former were outside the wheels whereas McIntosh put them inside, and the latter were fitted above the tender axleboxes instead of being underhung as Drummond had arranged them on his N.B. and Caledonian tenders.

By the last year of the century the new design was ready and went into production. Between May 1899 and July 1909, 96 of the new goods engines were built in the following batches: 812–838 at St Rollox in 1899, 829–848 by Neilson Reid 1899–1900, 849–863 by Sharp, Stewart 1900, 864–78 Dübs in 1900, 282–293 at St Rollox in 1899, 652–9, 661–5, 325–8 at St Rollox 1908–9. When new, and until 1918, Nos. 655 and 661 carried the numbers 423 and 460.

The boilers of these engines were similar to the 'Dunalastairs' but the cylinders were 18×26 in. which with 5-ft. wheels and 160 lb. per sq. in. pressure gave a tractive force of 20,169 lb. The engines weighed 45 tons 14 cwt. and the tenders 37 tons 18 cwt.

The slide valves were of phosphor-bronze and were placed vertically between the cylinders which had well proportioned ports. Lower production costs dictated the use of flat-faced pistons instead of the Stroudley/Drummond conical type. A feature which owed its origin to Stroudley was the means of operating the blower valve, fixed on the left side of the smokebox. From the valve a rod passed through the left handrail to a handle in the cab. The blower itself was a simple perforated pipe encircling the blast pipe top and connected to the valve which was supplied from the dome Gresham & Craven injectors and Vacuum Oil Co's lubricators (sight feed type) were among the cab fittings.

As in the case of the 'Dunalastair III' class the 'Jumbo' or '812' class 0–6–0 found favour in Belgium where a series of goods engines based on this design was introduced in 1906.

To enable passenger trains to be worked by some of these 0–6–0s a number were fitted with Westinghouse brakes and carriage warming apparatus, incorporating a Westinghouse pressure-reducing valve. Late in the steam era, after nationalization of the railways in 1948, the Westinghouse brakes were removed and Vacuum brakes substituted.

During the 1912 colliery strike and again in 1921 Nos. 285 and 292 were fitted with oil tanks and burners, in 1912 of the Holden type and in 1921 with Scarab equipment.

Under L.M.S. auspices when some effort towards standardization was being made 'Dunalastair' type boilers were constructed with slight alterations. Direct firebox crown stays replaced the girder stays, new smokeboxes the doors of which were secured by 'dogs' as in Midland practice, and strap-and-cotter type big ends in place of the marine type introduced by Drummond and carried on by McIntosh.

The first seventeen engines built at St Rollox were painted in the full Caledonian blue livery and were used on passenger services along the Clyde coast and on inter-urban and suburban services in the centres of high-density population. It is fortunate that one of these engines has been preserved and may be seen in all the glory of the original Caledonian colours in the Glasgow Museum of Transport.

The engines built at St Rollox in 1908–9 differed slightly from the remainder in several respects. A tendency to fracturing of the main frames near the horn gaps of the driving axle had been experienced and steps were taken to overcome this defect by strengthening the frames at this point. A liveliness in the springing resulting from the use of coil springs under the driving axleboxes was countered by replacing the coil springs by laminated springs. Thirdly the cabs were modelled on those of the 'Dunalastair III' and 'IV' classes. These engines were painted black and were intended for goods work only becoming known as the 'standard goods' engines of the Caledonian.

The mineral traffic of this company had been steadily growing. In the ten years from 1890 to the end of the nineteenth century the annual revenue from this source had risen from £882,511 to £1,244,617 and the mileage on this traffic from 6,270,629 to 7,350,140.*

A great need for more powerful locomotives to handle this type of traffic was met in July 1901 when the first of a new type of engine in Scotland emerged from St Rollox works. McIntosh wanted the maximum adhesion weight he could obtain and the highest tractive effort compatible with the dimensions he considered suitable to his purpose and within the dictates of the civil engineer. The result was a heavy eight-coupled locomotive with no carrying wheels, the total weight of 60 tons 12 cwt. 1 qr. being carried on the four axles. The spacing of the latter was unique: 8 ft. 6 in. from leading wheels to drivers, 5 ft. 4 in. from drivers to

* These mileages are for all goods and mineral traffic, the latter not being shown separately in the available statistics.

intermediates and 8 ft. 6 in. from intermediates to trailing wheels.

The cylinders were 21 × 26 in., between the frames and driving on the second axle. This second pair of wheels incorporated another unusual feature; the flanges were turned off. This was in itself a not uncommon feature but whereas in the L. & N.W.R. engines the Crewe practice was to have the intermediate wheels flangeless, in the Caledonian engines it was the driving wheels which were so treated. The wheels were 4 ft. 6 in. diameter and their springing included another novel arrangement. Whereas the three front axles had the usual underslung laminated springs the trailers were arranged differently. A pin through the tops of the trailing axle-boxes carried two screwed pins which projected into the cab. On the axlebox tops there were two spiral springs held in place and adjusted for compression by large nuts on a plate which was set transversely across the cab.

The boilers contained 275 × 1¾ in. tubes equal to 1,970 sq. ft. The firebox provided 138 sq. ft. in a total 2,108 sq. ft. The grate area was 25 sq. ft. and the working pressure 175 lb. per sq. in. The tractive effort was 31,584 lb. at 85 per cent BP and was the highest figure of any locomotive built in Scotland for one of the Scottish railways.

These ugly and cumbersome engines were by no means popular with the enginemen and must have been thoroughly disliked by the shed fitters also. Apart from the difficulty of changing springs, the slide valves were equally troublesome. They were on top of the cylinders and were driven through rocking shafts from the usual link motion, but to remove them for examination or renewal entailed removing the blast pipe and two large cover plates in the bottom of the smokebox and then rigging a block and tackle to lift the valves out.

In 1903 six more of these powerful engines were built at St Rollox but the design was not repeated thereafter. The running numbers allotted to the class were 600–7.

Six 0–8–0Ts the first of which, No. 492, came out of the company's works in November 1903 were followed before the end of the year by Nos. 493–5 and in 1904 by 496 and 497. Although these were an enlarged variant of the 0–6–0T already in service the boilers were not very different from the earlier 0–6–0 engines turned out by Drummond and Lambie. The total heating surface was only 1,189 sq. ft. made up of 1076·3 sq. ft. from 222 × 1¾ in. tubes and 112·7 sq. ft. of firebox area, while the grate area was no more than 19·3 sq. ft. This boiler was to provide steam for two inside cylinders, 19 × 26 in., driving on the second axle which was, as in the 0–8–0 tender engines, somewhat widely spaced from the leading axle. Again the driving wheels were flangeless and the total wheelbase was 19 ft.

Not particularly successful engines though capable of some quite good work on coal trains in the colliery areas, it might be wondered whether a McIntosh '812' class boiler would have improved them. Certainly the greater heating surface would have been advantageous but the increased weight may very well have put such a development out of court. As it was these engines scaled 62 tons 16 cwt. and the larger boilers might well have added at least a ton to the axleload thus still further limiting their sphere of usefulness.

For suburban workings, station shunting duties, and branch lines, the 0–4–4T was proving a useful engine. Originally introduced by Conner in 1873, Drummond had built more in the period 1884–91 and Lambie had added his quota, the '19' class for use on the Glasgow Central Railway, though they actually appeared after the designer's death. That Lambie was responsible for this design is debatable. Both these and the '3' class 4–4–0T had boilers containing 224 × 1⅝ in. tubes and, as already described, his 4–4–0 tender engines also employed this size of tube. The deduction to be drawn from this is that the drawings for the boilers, at least, of all three classes had been prepared under Smellie's supervision and possibly put into production before the change in command. McIntosh continued the construction by bringing out a further twelve condensing engines of this wheel arrangement in 1897 and ten more in 1899–1900. From then until 1922 ninety-four 0–4–4Ts were built in St Rollox works including four, Nos. 431–4, with 18¼ in. cylinders and a cast iron leading buffer beam. They were specially allotted to banking duties at Beattock and carried a pressure of 180 lb. per sq. in.

The majority of the 0–4–4Ts had 206 × 1¾ in. tubes bringing them into the range of Caledonian standards. Also included in this numerous class were engines which were specifically intended for, and did an immense amount of useful work on, the Cathcart

Circle Railway and the Balerno branch at Edinburgh. These were slightly smaller engines though conforming to the same general pattern. The cylinders were 17 × 24 in. and the overall weight 51 tons 2 cwt. They had boilers similar to the rest of the class.

Freight and dockyard shunting were catered for by 0–6–0Ts of the '29' class, the first appearing in 1895, following Lambie's '211' class. In these the cylinders were 18 × 26 in., the wheels 4 ft. 6 in. diameter and the boilers generally similar to the McIntosh 0–4–4T with 975 sq. ft. of tube surface and 110·9 sq. ft. of firebox. The grate was small, 17 sq. ft., and the pressure 150 lb. per sq. in.

The first nine engines were fitted with condensing apparatus for G.C.R. workings and weighed 49 tons 14 cwt. From 1898 until 1922 another 138 of these engines were built, all at St Rollox and non-condensing. Their weight was 47 tons 15 cwt. In other matters they were essentially the same as the first nine engines. The side tanks of the 0–6–0T, 0–4–4T and Lambie 4–4–0T are all recorded as holding 1,000 gallons and the coal spaces as being adequate for 2 tons except in the case of the 4–4–0T which held 2 tons 10 cwt. Known as the 'Beetlecrushers', McIntosh's last class of goods tank engines was brought out in 1911 when Nos. 498 and 499 were turned out. The cylinders were 17 × 24 in., wheels 4 ft. and the total heating surface 860 sq. ft. The grate was 14·25 sq. ft. and the working pressure 160 lb. per sq. in. These were the only outside cylinder engines McIntosh built and their by-name came from their weight, 47 tons 15 cwt. on the very short wheelbase of 10 ft. Pickersgill built a further 21 of these little dockyard engines between 1915 and 1922.

The year before McIntosh took charge of the Caledonian locomotive department the first 4–6–0 in Scotland, the 'Big Goods' of David Jones, had appeared on the Highland. It was not until 1902 that McIntosh turned his attention to the use of this wheel arrangement. In that year his '55' class made its debut on the Oban line for which it was intended and where the greater power and adhesion were required.

A tractive effort of 23,269 lb. was obtained from two 19 × 26 in. cylinders, 5 ft. coupled wheels and 175 lb. per sq. in. pressure. The boilers had the largest heating surface so far on the Caledonian, 1,905 sq. ft., from 105 sq. ft. of firebox and 1,800 sq. ft. from 275 × 1¾ in. tubes, while the grate area was 20·63 sq. ft.

Steam reverse was fitted, but whereas in other Caledonian engines it was fitted in the cab, on these it was on the main frame ahead of the driving (leading coupled) axle. Nos. 55–9 were built in 1902 and Nos. 51–4 in 1905. During the First World War No. 56 was loaned to the Highland Railway at a time when that company was short of power. This class was the second generation of 'Oban Bogie' and was also the first six-coupled express locomotive with inside cylinders on a British railway.

The following year saw the introduction of locomotives with the same wheel arrangement on the main line. In March 1903 two massive engines emerged from St Rollox works specially designed to cater for the ever increasing weight of trains between Carlisle and the north. The popularity of the West Coast Route was in no doubt and was by no means inhibited by the advertising campaigns of the two companies concerned. Even the competence of the 'Dunalastairs', all three types, was hardly equal to the tasks involved in negotiating the ascents of Beattock and Craigenhill banks with 300 ton trains and once again bigger and stronger locomotives were required. The answer was found in the two large 4–6–0s, Nos. 49 and 50, which McIntosh developed out of the Drummond basic designs.

The cylinders of the new engines were 21×26 in. and, as the diameter was too great to permit the slide valves being placed between them, the latter were placed above the cylinders. This had the effect of raising the boiler centre line to 8 ft. 6 in. above rail level and allowed a diameter of no less than 5 ft. An unusual arrangement of the barrel plates was embraced in this design; the barrel was in three rings, the front and rear rings being 5 ft. diameter while the middle ring, telescoped between them, was 4 ft. $10\frac{11}{16}$ in. diameter. This large barrel, 17 ft. 3 in. long between tubeplates, housed $257 \times 1\frac{3}{4}$ in. tubes and $13 \times 2\frac{1}{2}$ in. tubes, the larger diameter tubes being positioned at the bottom of the barrel. This was a good point and one adopted surprisingly infrequently in locomotive practice since it is the bottom rows of tubes which most usually get blocked up by products of combustion lifted from the firebed by the force of the draught on the fire. The tube heating surface was 2,178 sq. ft. which together with 145 sq. ft. of firebox surface produced a total of 2,323 sq. ft. whilst the grate area was 26 sq. ft. A further feature of the boiler was that the smokebox tubeplate was recessed into the barrel 9 in. thus enlarging the volume of the smokebox and at the same time shortening the tubes at their less effective end as regards evaporation. A complete departure from previous Caledonian practice was the adoption of the Crewe method of securing the firebox doorplate by riveting it with the flanges reversed from the normal practice. The working pressure of the new engines was 200 lb. per sq. in. and was controlled by four 4-in. lock-up valves mounted on the firebox.

The valves being on top of the cylinders necessitated indirect drive through rockers and the eccentric rods of the Stephenson link motion were therefore crossed. The frames, which were $1\frac{1}{8}$ in. thick, were very strongly braced with horizontal and vertical cross members and in addition to the normal spectacle plate another plate to the rear of it provided suspension points for pendulum links to the intermediate valve spindles.

Tenders for 5,000 gallons of water and 5 tons of coal ran on two four-wheel bogies and the springing was of the same type as was employed on the 'Dunalastair II' class using one inverted laminated spring for each pair of axleboxes.

These two engines were 65 ft. 6 in. in length, the wheelbase being 56 ft. $10\frac{1}{2}$ in. and there were no turntables of sufficient size to accommodate them. Accordingly, after detaching from their trains at Carlisle, they were required to run to the shed at Kingmoor where the tender was uncoupled, engine and tender turned separately and then coupled up again. It is said that the Carlisle fitting staff became so expert at this daily function that the operation could be completed in 20 minutes. It was not until 1906 that 70-ft. turntables were installed at a number of depots eliminating the need for such a performance. At the Glasgow end conditions were different; turning could be, and was, very easily accomplished by running the locomotive tender first round the Cathcart Circle line, as an alternative to going to Polmadie shed.

The space between the cylinder casting and the driving axle was considerably cramped, occupied as it was by main and auxiliary spectacle plates, rods and motion, so that when the next class of 4–6–0s was constructed in 1906 an attempt was made to improve matters by shortening the coupled wheelbase from 15 ft. to 14 ft. 8 in. but keeping the overall length the same so that the distance from the driving axle to the cylinder block, and similarly the length of the connecting rods, was increased by 4 in. It was not a great amount but it did make a slight improvement.

These new engines of 1906 were destined to become world famous. The first one, No. 903, was named *Cardean* after the estate of Edward Cox, Deputy Chairman of the company. The naming of locomotives on this line was not a common practice but No. 50 of the 1903 class was named *Sir James Thompson* in honour of the general manager who was the first Scottish railwayman to be awarded the accolade of knighthood.

Several modifications to the 1903 design were made in the new engines apart from that already mentioned. The cylinder diameter was reduced to 20 in., the stroke remained 26 in. and the coupled wheels 6 ft. 6 in. The length of the boiler and the number of tubes were reduced and the barrel comprised four rings. The internal diameter of the front ring was 4 ft. $10\frac{1}{4}$ in. and of the back ring 5 ft. $2\frac{3}{16}$ in. In this huge barrel were 242×2 in. tubes which, at 16 ft. 8 in. between tubeplates, gave 2117·5 sq. ft. of heating surface. This, added to the 124·25 sq. ft. of the firebox, made a total of 2265·75 sq. ft. The grate area was unaltered at 26 sq. ft. The ashpan was slightly modified and a back damper provided. The shallow pans of the '55' and '49' classes had a front damper only.

The staying of the firebox crown on the '49' class was by longitudinal girders slung from angle-irons inside the wrapper plate in the usual way, but in 903–907 there were four rows of sling stays at the front end of the firebox crown and seventeen rows of direct stays behind them. Since the inner box had a rounded top these direct stays were partly radial to the surface of the crown.

The coupled wheels were dished outwards in order to increase the distance between the inner faces of the wheel bosses. As a result the driving bearings were $10\frac{1}{2}$ in. long and $9\frac{1}{4}$ in. diameter. The leading and trailing journals were similar to those of the 'Dunalastair IV' 4–4–0, 12 in. long×8 in. diameter at the centre and $9\frac{1}{2}$ in. diameter at the ends.

Whilst the '49' class had scaled 70 tons 1 cwt., of which 53 tons 18 cwt. had rested on the coupled wheels, the 903 class weighed 73 tons with 54 tons 10 cwt. available for adhesion. *Cardean* went into regular service on the 2 p.m. 'Corridor' express Glasgow (Central) to London, returning from Carlisle with the corresponding down train. It was while working the latter on April 2, 1909, that the crank axle was fractured close to the left hand wheel seat. The failure was caused by inferior material and metal fatigue. A good account of this mishap is recorded in *Springburn Story* by John Thomas (David & Charles, 1964). In the Quintinshill disaster of May 22, 1915, No. 907 was so severely damaged as to be beyond repair and it was scrapped. These two unfortunate occurrences were more than made up for by the performance of *Cardean* and her sisters during their lifetime which lasted until 1927–30. In service their performance was good and, when tested against a L. & N.W.R. 'Experiment' class 4–6–0 between Carlisle and Preston and return in 1909, the power output recorded in the dynamometer car showed the superior qualities of the Scottish engine. On the climb from Penrith to Shap the 1 in 125 stretch was negotiated at a sustained 44 mp.h. with a drawbar pull of almost $4\frac{1}{4}$ tons and an equivalent DBHP of 1,390. On the return trip a train of 301 tons, 66 tons lighter than on the up run, breasted Greyrigg at 33 m.p.h. with 720 DBHP on the chart and Shap called for a figure of 1,000 DBHP. On neither run was assistance required on this steeply graded road.

In that same year, 1906, McIntosh brought out two other classes of 4–6–0s. One of these, the '918' class, of which five were built, were primarily a development of the Oban '55' class and were intended for express goods work. The other type was the '908' class sometimes described as a smaller edition of the 'Cardeans' and destined for Clyde Coast expresses and passenger work of secondary importance on the Glasgow-Perth-Aberdeen road.

Both classes incorporated a degree of standardization. The cylinders were 19×26 in. with slide valves vertically between. The boilers were the same diameter and the tube arrangement was the same but the lengths differed. In each case 242×2 in. tubes were employed, the lengths between tubeplates being as quoted in the technical press at the time, i.e. 13 ft. $6\frac{1}{2}$ in. in the goods engines and 14 ft. $11\frac{7}{16}$ in. in the passenger engines. The heating surfaces were thus 1715·8 sq. ft. and 1894·7 sq. ft. from tubes which, added to the 128 sq. ft. of the fireboxes, gave 1843·8 sq. ft. and 2025·3 sq. ft., the fireboxes and grates being the same in each class. Both *Engineering* and *The Engineer* of the period gave 1,895 sq. ft. and 2,050 sq. ft. for the same number, diameter and length of tubes, and these figures are not mathematically correct and must therefore be rejected in favour of the version quoted above.

As in the Oban engines the '918' class had 5-ft. coupled wheels and a working pressure of 175 lb. per sq. in. whilst the '908' class had 5 ft. 9 in. wheels and 180 lb. per sq. in. The wheelbase of the former was 20 ft. 11 in. and of the latter 23 ft., the overall lengths of engines and tenders being 57 ft. 6 in. and 58 ft. 11 in.

The tenders were standard 3,572 gallon tenders with coal space for $4\frac{1}{2}$ tons of coal and weighing 38 tons. The goods engine weighed 60 tons 8 cwt. and the passenger engines 64 tons so that the total weights were 98 tons 8 cwt. and 102 tons.

In all his work this engineer was outstandingly successful. None of his designs can be said to have been a failure. Some were inclined to be reluctant steamers but they were in the minority. First and foremost John Farquharson McIntosh was a running man. He knew what good enginemen wanted and he set out to give them just that. With the well-tried Drummonds as a foundation on which to build it might almost be said that success was assured from the start.

Of the five main Scottish companies the Caledonian, North British, Highland, and Glasgow & South Western were all developing Drummond design trends either as evolved by Dugald or his younger brother Peter, and in the case of the last two mentioned lines Peter's adaptation of the tradition was superimposed on the Allan and Stirling trends. Only the Great North of Scotland was unaffected by the Drummond dicta but remained satisfied with the small engines of the Cowan era, which was itself firmly based on Allan, and developed into its modern form by Manson and Johnson.

The next development of the steam locomotive was the fitting of superheaters and in this phase all five companies contributed in greater or lesser degree.

Chapter X

THE INTRODUCTION OF SUPERHEATING

As early as 1832 it was known that the raising of the temperature, or superheating of steam at a given pressure was beneficial and various experiments were carried out to find how best to adapt this development in thermodynamics to the steam locomotive.

A form of superheater was patented in 1839 by R. & W. Hawthorn and thirteen years later J. F. McConnell, locomotive superintendent of the Southern Division of the L. & N.W.R., tried out his first form of simple smokebox superheater. About the same time M. Moncheuil, on the Monterau & Troyes Railway in France, was experimenting.

These early attempts were by no means satisfactory and it was not until the late nineteenth century that a more practical solution to the problem appeared. This was the work of Dr Schmidt on the Prussian State Railways, whose first really successful superheater was produced in 1902. This was a smoketube superheater and was the basis on which future development was, for the most part, achieved. Early in the present century Cole in America, Cockerill in Belgium, among others, were experimenting, while in this country Mr (later Sir) J. A. F. Aspinall developed a form of apparatus on the Lancashire & Yorkshire Railway. This showed promise but the results obtained with it were insufficiently good to warrant its general introduction.

In Scotland the first railway engineer to make a positive move towards the adoption of superheating was J. F. McIntosh on the Caledonian Railway. In July 1910 No. 139, the last of the 'Dunalastair IV' class, was turned out of St Rollox works fitted with a modified Schmidt superheater. In August the same year it was fully tested on the 'Tourist' between Perth and Carlisle and came fully up to expectations.

On the G. & S.W.R. Manson did not lag far behind. He had two

4–6–0s, Nos. 128 and 129 built by N.B. Locomotive Co. in 1911, fitted with Schmidt apparatus. Both the North British and Highland Railways fitted out one engine with a Phoenix superheater in 1911 and 1912 respectively. Further north greater canniness was exhibited. The G.N.S.R. went no further than agreeing in 1912 to the fitting of Schmidt superheaters to one express passenger and one ordinary passenger locomotive.

Both companies experimenting with the Phoenix apparatus were dissatisfied with the results obtained and it was not long before the superheaters were removed. The successes with the Schmidt superheater in other parts of the country clearly pointed the direction to be taken and the N.B.R. was not slow to follow it. In 1912 the last two of Reid's 6 ft. 6 in. 4–4–0s first built in 1909, Nos. 363 and 400, were turned out from Cowlairs with Schmidt superheaters. On the Highland, however, despite the poor performance of the experimental apparatus it was retained in No. 141, *Ballindalloch Castle*, until the Cumming régime when it was removed in 1916.

While these experiments were proceeding the two pioneer companies were having considerable success. On the Caledonian No. 139 had been tested on trains of 220–235 tons and was showing a 16 per cent reduction in coal consumption per gross ton mile while the evaporation rate remained constant at 6·5 lb. per lb. of coal.

In this engine several changes had been made from the original design. The cylinder diameter was increased to 20 in., 8 in. piston valves with Schmidt trick ports were placed over the cylinders instead of between them as the flat valves had been, and as outside admission was the order of the day and the valves were driven through rocking shafts, the eccentric rods were crossed.

The boilers contained 163 small tubes 1¾ in. diameter and 24 superheater flue tubes 5 in. diameter; the length between tubeplates was 11 ft. 6 in. so that the tube heating surface was 1,220 sq. ft. This with 145 sq. ft. of firebox surface produced a total of 1,365 sq. ft. and the grate area was 21 sq. ft. as in the saturated engines.

In the early days of superheating it was customary practice for the boiler pressure to be lowered, in this case from 180 lb. per sq. in. to 165 lb. This was partly because for a boiler of given dimensions a larger volume of superheated steam could be produced than in a similar saturated boiler, while providing the same or greater haulage capacity in the engine as a whole. It was also partly due to a realization that the superheater engine and boiler would prove to be more expensive to maintain. A full appreciation of the potentialities of superheating was not gained or the advantages wholly exploited until very late in the steam era.

In order to prevent burning of the superheater elements when the regulator valve was closed dampers were provided. These were operated by a small steam cylinder on the side of the smokebox and controlled from the cab.

In 1911 four similar engines were built, five in 1912, seven in 1913 and a further five in 1914. Schmidt superheaters were fitted in the 1911 and 1912 batches, which were numbered 132–5 and 117–121, the last named being the ill-fated engine damaged beyond recovery in the Quintinshill smash of May 22, 1915. Nos. 122, 43–8 of 1913 and Nos. 40–2, and 123 of 1914, all had Robinson superheaters as developed by J. G. Robinson of the Great Central Railway. These too had damper gear but it was manually operated. The surface of the Robinson superheaters was 295 sq. ft. but the reduction from the 300 sq. ft. of the original Schmidt type was minimal and not inimical to performance. Save for No. 121 scrapped after Quintinshill and No. 133 cut up in 1946 all the rest of the class were taken into British Railways stock in 1948, the last to go under the hammer being 54441, old No. 132.

In addition to the tests already mentioned, No. 139 was indicated while working the 'Grampian Corridor Express' between Glasgow and Perth. The indicator diagrams reproduced in *The Caledonian Dunalastairs* by O. S. Nock are particularly good examples. Down to 32 per cent cut-off good fat diagrams are indicative of reasonably high mean effective pressures, and even at 21 per cent there is no sign of looping of the diagrams showing that the valve setting was very good indeed. No less than 940 IHP was developed at just over 50 m.p.h. at 21 per cent cut-off when passing Glenboig. The load was 205 tons. On Kinbuck bank a minimum speed of $21\frac{1}{2}$ m.p.h. produced 636 IHP.

The success of these engines led McIntosh to make slight modifications. The cylinders were enlarged to $20\frac{1}{4}$ in. diameter and the boiler pressure raised to 170 lb. per sq. in., an increase of over 5 per cent in tractive power.

The advantages of superheating were so evident in the work of these engines that McIntosh soon carried out the rebuilding of engines of the second and third generations of 'Dunalastairs'. Three of the former and two of the latter were superheated in 1914, followed by others in succeeding years. These engines were given 19½-in. cylinders and boilers with 159 × 1¾ in. and 18 × 5 in. tubes, 1,094·25 sq. ft. The superheater surfaces were 214 sq. ft. In performance these rebuilds showed themselves almost the equals of their larger sisters.

St Rollox was a busy works during the summer of 1911 for, in addition to the new superheater 4–4–0s, there were a number of goods and passenger tank engines built, as well as the rebuilding of the two big 4–6–0s, Nos. 49 and 50, and the 'Cardeans'. These seven engines were all given Schmidt superheaters with 24 elements. The cylinders were enlarged to 20¾ in. diameter and given Schmidt trick ported piston valves. The tubes were 2 in. and 5 in. diameter, 125 of the former in 49 and 50 and 132 in the '903' class; the heating surfaces were thus 1,509·25 sq. ft. and 1,666 sq. ft. respectively. The length between tubeplates of 49 and 50 was reduced to 16 ft. 8 in. by recessing the smokebox tubeplate into the barrel. The fireboxes were unaltered but the working pressures were dropped to 175 lb. per sq. in. An advantage of recessing the tubeplate into the barrel was that the smokebox could accommodate the superheater header without altering the shape and external appearance. The additional apparatus added 1 ton 5 cwt. to the weight. Scrapping of these seven engines commenced under L.M.S. auspices, the four 'Cardeans' being cut up in 1928–30, No. 907 had been broken up in the Quintinshill disaster, and 49 and 50 followed in 1933.

McIntosh's remaining 4–6–0 design was the big '179' class which first appeared in 1913 when five, Nos. 179–183, came from St Rollox followed in 1915 by Nos. 184–9. They were specially designed for the heavy goods trains, mostly long haul main line traffic. The cylinders were 19½ × 26 in. with 9 in. Schmidt piston valves with trick ports and indirect motion. The coupled wheels were 5 ft. 9 in. diameter. These engines had the Robinson superheater and a tube heating surface of 1,439 sq. ft. The fireboxes were the same as in the '908' and '918' classes, 128 sq. ft. and the grate area similar at 21 sq. ft. The superheater surface was 403 sq.

ft. and the pressure 170 lb. per sq. in. The cabs had two windows on each side, a departure new to the Caledonian with No. 917, the last of the '908s'. 51 tons 5 cwt. of the total weight of 68 tons 10 cwt. was available for adhesion.

The Caledonian only built four superheater 0–6–0s; these were Nos. 30–3 which came out in 1912. They also had $19\frac{1}{2}$ in. cylinders, Schmidt piston valves and 5-ft. wheels. The boilers were modelled on the '812' class with the centre line pitched 8 ft. 3 in. above rail level. $161 \times 1\frac{3}{4}$ in. and 21×5 in. tubes giving 780·5 sq. ft. and 290·8 sq. ft., with the 118·78 sq. ft. of the firebox, totalled 1,190·08 sq. ft. of surface. The grate area was 20·6 sq. ft. and the superheater surface 266·92 sq. ft. The working pressure was 160 lb. per sq. in. and the total engine weight 51 tons 2 cwt.

Though this class did some good work on Clyde Coast trains it was not perpetuated in its original form. There was rather a heavy overhang at the front end and this led McIntosh to modify the design by lengthening the frames at the leading end, fitting a two wheeled truck, making the engines the first 2–6–0s on a Scottish railway. The weight was increased to 54 tons 5 cwt. by this alteration.

McIntosh was now rapidly approaching retirement and in the autumn of 1913 when His Majesty King George V was returning to London from Balmoral McIntosh was called to the Royal day saloon to be invested with Membership of the Royal Victorian Order, a signal recognition for a locomotive engineer. In the following February he retired in his 73rd year.

Two proposed designs which might have influenced future locomotive development, but which actually only reached the drawing board stage, were a De Glehn compound 'Atlantic' and a 'Pacific' type. The compound proposal was dated 1905 when interest was centred on this method of propulsion and was a reasonably compact design but the 4–6–2 was no more than an enlarged 'Cardean'. Four cylinders 16×26 in. and the standard 6 ft. 6 in. coupled wheels would have given a tractive effort of only some 4,700 lb. more than the 'Cardeans' at 85 per cent of 180 lb. pressure. The boiler was to have had $135 \times 2\frac{1}{4}$ in. and 24×5 in. tubes equal to 1,750 sq. ft. The superheater surface was put up to 526 sq. ft. and the firebox 158 sq. ft. while the grate was to be the largest so far in Scotland, 37 sq. ft. These dimensions do not seem to justify the greater expense of manufacture of an engine weighing about 90 tons, some 16 tons more than the '903s' but having only the same adhesion weight.

The second venture in superheating was Manson's 4-6-0s on the G. & S.W. In July 1911 the North British Locomotive Co. built Nos. 128 and 129 based on the 1903 engines and incorporating Schmidt superheaters. The cylinders were 21×26 in. and the coupled wheels 6 ft. 6 in. As usual the pressure was dropped to 160 lb. per sq. in. In service these engines proved to be good, particularly No. 129. This engine also distinguished itself by running as a 4-4-2 for a while in 1927 because of a broken trailing portion of a coupling rod. This was at the time when it was working the 1.50 p.m. Glasgow (Central) to Carlisle *via* Carstairs as a relief to the Midday Scot and taking the Edinburgh portion forward from Strawfrank Junction. The return working was the 6.34 p.m. from Carlisle which was the 12 noon from St Pancras. Out *via* Beattock and home *via* Polwhap.

In length and diameter these two boilers were similar to the '381' class but had $21 \times 5\frac{1}{4}$ in. and 119×2 in. tubes giving 450 sq. ft. and 980 sq. ft. which, with the 130 sq. ft. of the firebox, produced a total heating surface of 1,560 sq. ft. The grate had an area of 24·5 sq. ft. and the Schmidt superheater provided 445 sq. ft.

When new, No. 129 was fitted with a Weir Feed Water Heater and Pump, the condenser unit being mounted on top of the boiler between the dome and the smokebox.

These were Manson's last engines. He retired at the end of 1911, his successor being Peter Drummond. Manson was 67 years of age and Drummond five years his junior. In retirement Manson, sometimes described as the 'Admirable Manson' because of his innate fairness towards his staff coupled with a strong sense of discipline and kindliness of manner, lived to the ripe old age of 90. During this time he saw much of his life's work swept away first by Drummond and later by Whitelegg before the wholesale scrapping of G. & S.W. locomotives by the L.M.S.

One of the first shocks suffered by the South Western when Drummond took charge was the change over of the driver's position from the right side of the footplate to the left side in common with Drummond practice. He then proceeded to produce new locomotives which combined massiveness with other features common to both brothers' designs.

First, in 1913 came fifteen very large o–6–o goods engines. They had $19\frac{1}{2} \times 26$ in. inside cylinders and 5-ft. wheels. The boilers were 11 ft. $8\frac{5}{8}$ in. between tubeplates with $295 \times 1\frac{3}{4}$ in. tubes while the fireboxes were of 147 sq. ft. area and the grates 26·2 sq. ft. The working pressure was 180 lb. per sq. in. so that the tractive effort at 85 per cent BP was 25,210 lb. Their weight was 57 tons 14 cwt. which with the 45 tons 6 cwt. of the tenders made them the heaviest and most powerful o–6–os in the country when built. In two respects these engines incorporated novelties so far as the G. & S.W.R., or Scotland for that matter, was concerned. Their designer, like the elder Drummond, was no believer in even comparatively low degree superheat. Instead both men considered adequate improvement would be effected by raising the temperature of the steam by no more than about 20° Fahrenheit. This result was obtained by fixing in the smokebox a pair of boxes in front of the tubeplate and on either side of the blast pipe. These boxes contained a number of tubes slightly larger in diameter than the boiler tubes and exactly opposite them. Steam entered the boxes from a 'Tee' joint on the smokebox tubeplate, was further heated by transferrence from the tubes which were surrounded by the steam, and thence passed to the steam chests.

The second piece of apparatus which was new to the system was the feed water pumps instead of injectors, designed to work with hot water. Some of the exhaust steam from the cylinders together with the exhaust from the feed pumps beneath the footplate was passed through a nest of some sixty tubes in the well of the tender tank. The rear ends of the tubes were turned down through the tank bottom so as to allow condensed exhaust steam to drain on to the permanent way. Keeping the tender tanks free from leakage caused by vibration of the tubes was virtually impossible, and the tenders were like water carts. These engines would steam well but on a coal consumption of 90–100 lb. per mile. They were sluggish, being unable to get rid of their exhaust freely. The class took its designation from the first built, No. 279, but was better known as 'The Pumpers'. On the Long Road Goods they loaded up to fifty wagons but needed assistance if time was to be kept on the climb from Hurlford to Polwhap. Drummond favoured marine type big ends and they gave endless trouble, not only on these but on other G. & S.W. engines similarly fitted.

1913 also saw a class of 4–4–0s built by N.B. Locomotive Co. to replace the Manson 'Greenock Bogies'. Six extremely handsome engines, Nos. 131–6 based on the L. & S.W.R. '463' class, had similar cylinders to the 0–6–0s and 6-ft. coupled wheels. The boiler barrel was slightly longer and contained ten additional tubes. Heating surfaces were therefore: tubes 1,736 sq. ft., firebox 148 sq. ft. The grate was increased to 27·6 sq. ft. The boiler pressure was 180 lb. per sq. in. and steam driers were fitted. These engines had the usual Drummond features and Westinghouse brakes. Piston valves above the cylinders were driven by Walscheart valve gear between the cranks. At first the single eccentrics of this gear had no liners and almost at once lubrication troubles were set up, the oil passages becoming blocked with abraded cast iron. The cure was the provision of white metal liners. Overheating of bogie axle-boxes and the marine big ends continued to give trouble. An odd experiment was carried out on No. 136. This engine was fitted with a grate which was humped in the centre. The arrangement may have improved combustion but fire cleaning must have been a very difficult job.

On the road these engines tended to roll badly when running at speed. South Western practice was generally to avoid pounding uphill and then to run like the wind downhill. This was not always the best policy with the '131s' which were liable to set up a bad roll especially if a soft spot should be found in the track.

The '137' class which came from Kilmarnock in 1914–15 was a great improvement. They were numbered 137–140, 151, 152, and differed from the foregoing class in having superheaters of the Schmidt type, and in retaining their 180 lb. pressure. There were 22 large and 187 small tubes of the standard 5 in. and 1¾ in. providing 1,444 sq. ft. surface. Firebox and grate were as before and the superheater surface was 330 sq. ft. Drummond had at last realized that full superheat could no longer be deferred. Feed water heating was applied but this time by the Weir system, with hot water injectors. All this additional equipment added considerably to the weight and, as there was no weighbridge capable of taking the axle weight of over 20 tons, estimates have always been accepted. The '131' class scaled 21 tons 17 cwt. on the bogie and 39 tons 17 cwt. on the coupled axles. It is evident that the '137' class must have carried some 22 tons on the bogie and the remaining 42 tons of the estimated 64 tons on the coupled wheels. These engines suffered all the ailments of the former Drummonds but were excellent hill climbers so that they did not require to be raced downhill to recover time lost on the climb and the benefit of the superheaters was apparent in the economies obtained. On the Port Road the Stranraer men were burning no more than 25 lb. of coal per mile.

It was at this period that World War I broke out. Drummond had a new goods engine on the drawing-board. To accommodate the additional weight of the superheater and avoid excessive fore-end overhang these were to be 2–6–0s. Once again N.B. Loco. Co. was given the order and the old Dübs establishment at Queens Park turned out the order of 11 engines. The first six were numbered 403–8 but were renumbered before going into traffic. They were followed by Nos. 16, 61, 116, 117 and 121, the first named, 16, giving the name of the class, though they were usually known as the 'Austrian Goods' or more commonly, 'The big Austrians'. Their only claim to association with Middle Europe is that in their construction use was made of such materials as were suitable left from a contract for Austria cancelled by the outbreak of hostilities. Again the cylinders and coupled wheels were $19\frac{1}{2} \times 26$ in. and 5 ft. The boiler barrel was similar to that of the 0–6–0s but contained the same tubing arrangements as the '137' class. The heating surfaces were tubes 1,344 sq. ft. while the firebox was 147 sq. ft. as in the 0–6–0s, and the grate was again 26·2 sq. ft. Instead of Schmidt superheaters this time Drummond used the Robinson Short Return Loop pattern with draught retarders.

The big 0–6–4Ts on the Highland Railway were doing good work and this prompted Drummond to modify the design to make a suitable 0–6–2T for the steeply graded Ayrshire mineral lines with their heavy traffic in coal and ironstone. Six of these were built by N.B. Loco. Co. in 1915–16 followed by a further twelve in 1917. Once again 5-ft. coupled wheels were used, this time in conjunction with $18\frac{1}{2} \times 26$ in. cylinders. Heating surfaces were: tubes 1,144 sq. ft., firebox 118 sq. ft., total 1,262 sq. ft. There was no superheater and the grate and pressure were 22·4 sq. ft. and 180 lb. per sq. in. respectively. These well proportioned engines had a good turn of speed but there is no record of them having been used on passenger trains at any time. They were all taken over by the L.M.S.

and lasted until 1937-8. At the time of Drummond's death another ten were on order from the same builders. Whitelegg made some modifications and increased the tank capacity to 1,910 gallons and restored the driver's position to the right side. They came out in 1919 and were numbered 1-10 in the renumbering scheme initiated by Whitelegg in that year. Both batches were widely scattered in later years and several survived until after nationalization; No. 16905, old No. 6, being withdrawn in April 1948, was the last G. & S.W.R. engine to remain in service.

Drummond designed one other class of engine for the South Western, again a tank engine, the No. 5 class which appeared in 1917. These were smaller engines weighing only 40 tons and having 17 × 22 in. cylinders and 4 ft. 2 in. wheels. The heating surfaces were also small: tubes 758 sq. ft., firebox 118 sq. ft. totalling 876 sq. ft. The grate was 17 sq. ft. and the pressure 160 lb. The N.B. Loco. Co. built these three engines which were numbers 5, 7 and 9. They all passed into L.M.S. stock and were withdrawn in 1932-4. These 0-6-0Ts were specially for light permanent way and sharply curved branches where they replaced the old Stirling 0-4-0 tender engines which were fast wearing out at nearly 50 years of age. The driving wheels were flangeless, the cylinders were outside and the valve gear was Walscheart. After withdrawal Nos. 7 and 9, by then 16378 and 16379, were sold to Carlton Collieries. Subsequently in 1966 old No. 9 was refurbished in the old livery and preserved in the Glasgow Transport Museum.

A Drummond dream never realized was a 4-6-0 based on the elder brother's L. & S.W.R. 'Paddleboxes'. Another project which remained at the drawing office stage was a 4-4-2T envisaged in 1915. This was to have had 19½-in. cylinders, 5 ft. 9 in. coupled wheels and a heating surface of 1,233·8 sq. ft., tubes 134·2 sq. ft., firebox 290·4 sq. ft. superheater and a 26·25 sq. ft. grate. At a pressure of 170 lb., a tractive effort of just over 20,000 lb., and with a tank capacity of 1,920 gallons a reasonable range of activity was ensured. They could have proved useful engines on suburban and coast trains.

Drummond had commenced his association with this company heartily disliked by his staff. After his initial efforts which produced the troublesome earlier designs his engines became better liked and even appreciated by the men handling them. The turning point was the adoption of good superheaters instead of the driers he had started out with. Throughout his career, both on the Highland and on the South Western, Peter Drummond seems to have been reliant on the work of the elder brother on the other South Western line. Peter died, in service, on June 30, 1918, at the age of 68, and was succeeded by R. H. Whitelegg.

At Cowlairs W. P. Reid had fitted a Phoenix superheater in the smokebox of No. 897, one of the 1909 'Scott' class 4–4–0s, and in No. 329, a Reid 5-ft. goods built in 1906.

This was a cumbersome apparatus which occupied a considerable space in the smokebox. Two headers were placed, one on either side of the blast pipe, near the bottom of the box. One communicated directly with the steam chest. Between these headers was a large number of tubes so arranged that the steam, which was taken from the boiler via the main steam pipe to the other header, passed from header to header five times before reaching the steam chest. To ensure a free passage for the exhaust special castings surrounded the chimney petticoat pipe, and a baffle plate behind the blast pipe deflected the gases from the boiler tubes amongst the tubes of the superheater. It was claimed that the temperature of the steam was raised to 500° Fahrenheit or even 550° Fahrenheit, and that trouble from leakage of the many tube joints did not arise. The nature of the construction necessitated the chimney being placed very far forward on the smokebox, giving any engine so fitted a most unsightly and unbalanced appearance. Whatever the merits of this form of superheater it found no favour in this country.

There had been much discussion of the subject of superheating in 1910–11 and as a result of the poor performance of the Phoenix apparatus the last two engines of the 1911 order for 'Scott' class engines, Nos. 363 and 400, appeared in 1912 with Schmidt superheaters. They had 8-in. piston valves above the cylinders driven through rocking shafts and were fitted with steam operated dampers which were soon removed. This was the start of the 'Superheater Scott' class of which a further twenty-five engines were built in 1914–20. During this period the 'Glen' class made its début. These were a superheated version of the 6-ft. 'Intermediates' of 1909–10 and although found on various parts of the system were especially associated with the West Highland line. The cylinders were 20 × 26 in. with 10-in. piston valves and Stephenson link motion. The boiler diameter was 5 ft. and its length 11 ft. 4 in. From the start they had 124×1¾ in. and 22×5 in. tubes which, with the 139·7 sq. ft. of the firebox, gave a total heating surface of 1,153·14 sq. ft. The first five engines built in 1913 had Schmidt superheaters with manually operated dampers which were soon removed. The remaining twenty-seven engines had Robinson superheaters with draught retarders. The heating surface of these superheaters was 192·92 sq. ft. and the use of the short return loop elements appears to have been dictated by the theory that the greater heat transferrence took place in the rear half of the element and that the portion between the dome and the smokebox was comparatively useless. Although many companies were having satisfactory results with long elements as used in the Schmidts, the Robinson Short Loop elements were gaining popularity in Scotland and became the standard on the N.B. and G.N.S. Railways.

The 'Glens' weighed 57 tons 4 cwt., of which 37 tons 3 cwt. was on the coupled axles, with a maximum axle load of 19 tons 2 cwt. The 46 tons 13 cwt. tenders held 4,235 gallons and 7 tons of coal. The grate area was 21·13 sq. ft. and the boiler pressure 180 lb. per sq. in. They had the respectable tractive effort of 22,100 lb. The numbers carried by the five Schmidt engines of 1913 were 307 and 405–8, and of the Robinson engines of the same year: 149, 221, 256, 258 and 266. The remainder were Nos. 100, 291, 298, 153 and 241 built in 1917, 242, 270, 278, 281, and 287 in 1919, and Nos. 490, 492–6, 502–5, 534 and 535 in 1920. The class was scrapped in the years 1946–62 but No. 256 *Glen Douglas* has been preserved in the Glasgow Transport Museum.

The 6 ft. 6 in. 'Superheater Scotts' came from Cowlairs in batches, Nos. 409–423 in 1914, 424–8 in 1915 and 497–501 in 1920. Robinson superheaters were fitted, and like the 'Glens' they had 10-in. piston valves. These larger diameter valves undoubtedly contributed to the free running of these engines. The original boilers had 152×1⅞ in. and 18×5½ in. tubes giving 1,166·9 sq. ft. of heating surface while the firebox had 139·7 sq. ft. and the grate area was 21·13 sq. ft. In 1921 they were given a 24-element super-heater with 5-in. flues and the number of small tubes was reduced to 136 but still 1⅞ in. diameter; in 1924, after the Amalgamation, something near a standard was reached with 124×1¾ in. and 22×5 in. tubes.

A somewhat similar development period occurred with the 0–6–0 goods engines. The year 1914 saw the introduction of the superheated version of the 1906 engines. The cylinders were 1 in. larger diameter, $19\frac{1}{2}$ in., and had $9\frac{1}{2}$-in. piston valves with inside admission. The valves were above the cylinders and necessitated the boiler centre line being pitched $6\frac{1}{4}$ in. higher than in the earlier engines. The tubing of the boilers was $175 \times 1\frac{7}{8}$ in. and 24×5 in. giving 1,265·5 sq. ft. with 142 sq. ft. of firebox surface and a 19·8 sq. ft. grate. The pressure was 175 lb. per sq. in. In passing it may be noted that with these and the passenger engines there was no reduction in the working pressure when superheaters were fitted but immediately after the L.N.E. assumed control the figure was raised to 180 lb. in those cases where it had been lower. In 1924 these and the 1906 engines had a modified arrangement of tubes using $153 \times 1\frac{13}{16}$ in., an unusual size, for the small tubes, and 22×5 in. for the large flues. These were later changed to $155 \times 1\frac{3}{4}$ in. and $22 \times 5\frac{1}{4}$ in.

In all there were 104 of these superheater goods engines built as follows: five in 1914–15, ten in 1915, five in 1916 and 1918 and a further ten in 1921, all at Cowlairs and all carrying 165 lb. per sq. in. pressure. The N.B. Loco. Co. at their Atlas works built thirty-four in 1918–19 with the same pressure, twenty in 1919–20 with 175 lb. and a final fifteen at 165 lb. in 1920–1. The weight of the engines was 54 tons 14 cwt. with 20 tons 6 cwt. on the driving axle. The leading axle was carrying rather more weight than was satisfactory and heating troubles were rife. The L.N.E. took a hand in 1927 and carried out some experiments with various weight distributions finally transferring 18 cwt. from the leaders to the drivers.

In June 1917 one of this class was tested against a N.E.R. 2-cylinder 0–8–0 as a result of which, being satisfied with the performance of their own 0–6–0, the N.B. board decided against building eight-coupled freight engines. However the question arose again a few years later and some tests were carried out between Bridge of Earn and Glenfarg. The engines involved were a Great Western 2–8–0 No. 2804 and N.B.R. No. 46, one of the 1920 0–6–0s with the pressure raised to 180 lb. per sq. in. On January 12, 1921, the latter engine stalled on the 1 in 75 bank with a load of 477 tons behind the tender. The G.W.R. engine took 590 tons successfully but stalled with 683 tons. On the second run the

weather conditions were bad. A third test was made in the following August, when N.E.R. 3-cylinder 0–8–0 No. 903 negotiated this
fearsome bank of some 5½ miles at 1 in 75–74½ with loads of 617,
703 and 755 tons. By this time however the Amalgamation of 1923
was imminent and there was no justification to spend money on
new building.

Only one class of tank engine was superheated in N.B.R. days.
This was an excellent 4–4–2T for suburban services mainly in the
Glasgow and Edinburgh areas. With 19×26 in. cylinders, 5 ft. 9
in. coupled wheels and 165 lb. per sq. in. pressure their tractive
effort was 19,078 lb. The boilers had 142×1¾ in. and 18×5 in.
tubes equal to 930 sq. ft. The firebox heating surface was only 95
sq. ft. and the grate was small, 16·5 sq. ft. Scaling 72 tons 10 cwt.
with 37 tons 5 cwt. available for adhesion they carried 2,080 gallons
of water and 4½ tons of coal. The N.B. Loco. Co. built them all:
Nos. 438–442 in 1915, 443–452 in 1916 and 511–16 in 1921.

An experiment with a type of smokebox steam drier was made
by Reid in 1905–7. An American invention developed by the New
Century Engine Co. was fitted to 0–6–0 No. 656 in 1905 and to
'Atlantic' No. 874 in 1907. No data has survived of this apparatus
which consisted of a means of pumping a mixture of air, compressed by crosshead driven water-cooled pumps to a smokebox
steam drier and thence to the cylinders. The 0–6–0 was given some
road tests and No. 874 was tested on a form of dynamometer at
Beardmore's works at Dalmuir.

The Great North of Scotland Railway was the last in Scotland
to adopt superheating. Whether this was due to the innate canniness of the governing body or to the fact that the company was
carrying on its business in a very brisk manner with the locomotives it possessed cannot be determined with any certainty
today. At the time the other companies were fitting their locomotives with the new equipment the Great North board still discussed the possibilities—and the cost—of fitting engines with
superheaters.

The first steps seem to have been taken in 1912 when the
directors ordered that one express passenger and one ordinary
passenger engine be fitted with the Schmidt apparatus. Accordingly No. 77, a Manson 6 ft. 6½ in. 'Q' class was fitted in October
1913 and No. 74, a Manson 6 ft. 0½ in. 'O' class was equipped in

April 1916. Satisfactory results were obtained and in October 1916 the board sanctioned the fitting of three more engines. By this time the Robinson superheater had been developed and had gained popularity and was used in the following engines in 1917: No. 12 ('P' class) in May, No. 14 ('P' class) in October and No. 75 ('Q' class) in July.

After this demonstration four more 'O' class engines were recommended for conversion in 1918. These were Nos. 17, 18, 72, and 73 which were fitted with Robinson superheaters. No. 74, which originally had Schmidt apparatus, had this changed for a Robinson superheater sometime after 1923 but No. 77 retained its original superheater until it was withdrawn in 1937. All three classes of engine had similar boilers and the tubes numbered $104 \times 1\frac{3}{4}$ in. and $18 \times 4\frac{3}{4}$ in. The tube heating surface was thus only 754 sq. ft. and, together with the 106 sq. ft. of the firebox, made a total of no more than 860 sq. ft. Pickersgill had introduced superheating to the G.N.S. but it was left to his successor T. E. Heywood to develop it.

It is recorded in the minutes of the Highland Railway Locomotive Committee dated November 24, 1903, that Frederic Godfrey Smith be appointed works manager at a salary of £250 per annum. This was an appointment which was to be of some consequence to the company for when Drummond moved south in February 1912 Smith was appointed in his place. His salary was £500, precisely the same as Stroudley's 43 years before. Thomas Brown, locomotive superintendent at Berwick N.B.R., was given the post of assistant to Smith at £200 p.a.

Smith was born in 1872 and served a pupilage under T. W. Worsdell on the N.E.R. at Gateshead. Later he gained further experience in the running department and spent four years with various electrical firms before returning to railway work at Inverness.

The first additions to the locomotive stock were four 'Castle' class 4–6–0 carrying 180 lb. per sq. in. and having extended smokeboxes. The chimneys had capuchons and the big ends were fitted with solid bushes instead of split brasses and cotters. Built by N.B. Loco. Co. at Queen's Park, they carried the numbers 26–8 and 43, and they were named.

The Highland Railway was proving to be one of the most

important strategic arteries in the country during World War I, providing the direct and only rail link to the far north whereby the Royal Navy at Scapa Flow was served *via* Thurso. The exceptionally heavy traffic, now an all the year round feature instead of seasonal as before the war, played havoc with the condition of the rolling stock, in particular the locomotives. The requirements of the operating department and the shortage of materials meant the deferment of repairs until engines became complete failures. In August 1915 Smith, Alexander Newlands the chief engineer, and Robert Park the general manager met the Railway Executive Committee at Perth with a view to obtaining some engines on loan from other companies. Of 152 Highland engines, 50 were urgently in need of heavy repairs but were still at work, 50 were withdrawn from service for the same reason and two had been scrapped. Arrangements were made for the loan of twenty locomotives. Some easement of the situation was thus achieved and it was further anticipated that more aid would be obtained when the six new 4–6–0s of Smith's design arrived from the maker's works at Newcastle upon Tyne. Hawthorn Leslie & Co. delivered the first two engines at Perth in September 1915 but they were found to be too heavy for certain of the bridges and produced too great a hammer blow on the track; they were slightly out of gauge by the Highland loading gauge, being too tall at chimney, dome cover, safety valves and whistle, and wide to gauge at leading, cab and tender steps and the outside lower quadrant of the cylinders and the drain cocks. Newlands prohibited them from running on the Highland metals and the resultant row ended with Smith being asked for his resignation. The offending engines were sold to the Caledonian for £5,400 each; they had cost the Highland £4,920 each!

These 4–6–0s were examples of good design at the time of their appearance. The cylinders were 21 × 28 in. and outside the frames. The bogie and coupled wheels were 3 ft. 3 in. and 6 ft. diameter respectively, and the bogie centre pin was equidistant from both axles, the wheelbase being 6 ft. 6 in. From bogie centre to leading coupled axle was 8 ft. 10½ in., thence to driving axle 6 ft. 3 in. and driving axle to trailing axle 8 ft., totalling 26 ft. 4½ in. The boiler was in three rings, the outside diameter of the centre, smallest, ring being 5 ft. 1$\frac{13}{16}$ in. Length between tubeplates 14 ft. 10$\frac{7}{8}$ in. so that the 24 × 5 in. flues and 125 × 2 in. small tubes gave 1,460 sq. ft. which, with the 139·6 sq. ft. of the firebox, produced a total of 1,599·6 sq. ft. The superheater was a Robinson type with short loop elements having an area of 350 sq. ft. The grate area was 25·3 sq. ft. and the front set of firebars was hinged to operate as a drop grate manually controlled from a handwheel in the cab. Although the boilers were designed for a working pressure of 180 lb. per sq. in. there is no record of them having been pressed to more than 160 lb. A well-designed Walschearts gear ensured a good distribution of steam through 10-in. inside admission piston valves whose lap was 1 in., lead $\frac{3}{16}$ in. and maximum travel $4\frac{7}{8}$ in., all of which particulars were suitable for the type of work and road for which the engines were intended.

A steam reversing mechanism was mounted horizontally inside the right hand frame between the leading and driving wheels. All coupled wheel bearing springs were laminated, as were the tender springs, and the bogie suspension was by independent coil springs for each axlebox. The bogie side play was $1\frac{1}{2}$ in. each way, centring being assisted by laminated springs in the cradle. The brake gear was somewhat complicated. A 21-in. vacuum cylinder between the leading and driving axles applied blocks to the back of the driving and trailing wheels, and a similar cylinder beneath the footplate applied blocks to the front of the leaders and trailers. For handling 'foreign' stock Westinghouse brakes were also fitted. Smith's own design of feed water heater was another adjunct and the boiler was fed by two Davies & Metcalf no. 11 injectors. A Wakefield no. 1 ten feed mechanical lubricator for the steam chests and cylinders was mounted on the right hand running plate. Auxiliary oil boxes with worsted trimmings fed the coupled wheel axlebox crowns and the hornplates and Armstrong oiler pads were fitted in all axlebox underkeeps.

The engine weight in working order was 72 tons 6 cwt. 3 qr., of which 19 tons 6 cwt. rested on the bogie, 17 tons 15 cwt. 1 qr. on the leaders, 17 tons 15 cwt. on the drivers and 17 tons 10 cwt. 2 qr. on the trailers. The six wheel tenders had a capacity for 4,000 gallons of water and $6\frac{1}{2}$ tons of coal weighing in all 49 tons 3 cwt. 3 qr. This, with the engine, made a total weight of 121 tons 10 cwt. 2 qr. The leading and trailing tender axles were given $\frac{1}{4}$ in. side-play and the middle axle had $\frac{3}{8}$ in. The Highland intended to name these engines after territorial rivers, Nos. 70–5 being called *River Ness, Spey, Tay, Findhorn, Garry* and *Tummel*. Only the first two carried their names.

When the Caledonian received these engines they were numbered 938–943 in the Caledonian series. The drop grates and feed heaters were removed and the cab roofs fitted with supporting columns.

During 1915 some tests had been carried out with 'Dunalastair IV' No. 136 fitted with feed heating apparatus designed by G. & J. Weir Ltd. Cathcart, Glasgow. A similar engine with an injector fed boiler ran on alternate days with No. 136 both making three return trips between Glasgow and Carlisle, 204·62 miles, with loads between 201 and 231 tons behind the former and 240–272 tons behind No. 136. The injector fed engine consumed an average of 51·8 lb. coal per mile for an evaporation rate of 6·89 lb. water per lb. coal. while the pump fed engine averaged 52·82 lb. per mile for 7·52 lb. water per lb. coal. The coal per ton mile figures were 0·235 lb. and 0·205 lb. in favour of the pump and heater. Thus the Weir apparatus showed an economy of 12·72 per cent in coal consumption and 9·14 per cent improvement in the evaporation rate.

The Smith heater used on some of the Highland locomotives was a means of using injectors to feed the boiler after the feed water temperature had been suitably raised. A chamber between the frames contained a nest of tubes around which exhaust steam from the cylinders, taken from the base of the blast pipe, was allowed to circulate. The feed from the injector was passed through the tubes and its temperature thereby raised. The feed was then taken past the boiler clack valve to another heater in the smokebox arranged after the manner of the Drummond driers. From this heater the feed was finally delivered to the boiler just behind the smokebox tubeplate.

Tests carried out in 1914 with 4–4–0 No. 62, one of the 'Big Ben' class, showed that the heating apparatus produced a saving of 22·71 lb. coal per mile, or 33·25 per cent. A second engine tested was No. 65, 0–6–4T, which before being fitted with the Smith apparatus was consuming 74·15 lb. per mile. After fitting the heater the consumption fell to 51·77 lb. per mile, a saving of 22·38 lb. per mile, or 30·18 per cent. In terms of money this represented a saving of approximately £9 10s per 1,000 miles run.

On the G. & S.W.R. James Manson made some tests in the autumn of 1911 with three of his 4–6–0 locomotives. No. 123, saturated, and No. 128, superheated, were run in comparison with No. 129 which was not only superheated but had the Weir feed water heater and pump as well. Each engine ran from Glasgow to Carlisle and return, 231 miles, with a trailing load of 242 tons, at an average speed of 50 m.p.h. The coal consumption figures were rather startling. No. 123 burned 54·46 lb. per mile, No. 128 44·76 lb. per mile, while No. 129 without the Weir apparatus in operation was similar at 44·28 lb. per mile, but with the heater and pump in operation the consumption was down to 40·24 lb. per mile. The evaporation rate corresponding to the latter figure was 7·75 lb. water per lb. coal. Thus the Weir apparatus showed an economy of 26 per cent coal and 22 per cent water over No. 123 and 8½ per cent saving in coal over the injector fed superheater engine. In the latter case No. 129 showed a greater water consumption than No. 128 by 1½ per cent. During the tests the average feed water temperature was 175° Fahrenheit and the superheater temperature ranged from 600° Fahrenheit to 650° Fahrenheit.

The G.N.S.R. does not appear to have experimented in this direction but the N.B.R. fitted a 'Scott' class 4–4–0, No. 359, with Weir heater and pump in 1915. This was removed in 1918 and no data appear to have been left.

Although the various devices for ensuring a measure of economy in the consumption of coal and water did show that some reasonable savings could be effected, the fact that few of the special fittings survived for any great length of time is perhaps explained by the cost of their maintenance. So often in the course of locomotive history claims, sometimes reasonable, often extravagant, have been made for economies obtainable by the use of special equipment. If these were commercially viable, and did not incur heavy expenditure of time and money in repairs and maintenance, and the stocking of spare parts, then it is reasonable to suppose more use would have been made of them, and even greater economies effected.

When he arrived at St Rollox, Pickersgill found a fleet of excellent 4–4–0s and some good 4–6–0s both types having benefitted from the application of superheaters. Shortly after the start of World

War I more locomotives were necessary and Pickersgill's choice was the ubiquitous 4–4–0. Instead of continuing the successful line of superheated 'Dunalastair IVs' he set out to improve on the design. This so called improvement consisted mainly of making everything heavier and more robust than before. The cylinder centres were 3 in. closer permitting a substantial increase in the size of the coupled axle bearings, as shown in the table on p. 193. The frames were made $1\frac{1}{4}$ in. instead of $1\frac{1}{16}$ in. thick. The boilers were generally similar but suffered from some modification. The small tubes were reduced in number to 157 but remained the same diameter. The large tubes were similar in number and size to the McIntosh standards but Robinson Short Loop elements were used and gave a surface of only 200 sq. ft. This, with the 1,185 sq. ft. of the tube surface and 144 sq. ft. of firebox, produced a combined total 131 sq. ft. less than the 'Dunalastair IV'. The degree of superheat obtained was probably little, if at all, less than in the McIntosh engines because of the greater efficiency of the rear part of the elements as compared with the front portion.

It was about this time that the Association of Railway Locomotive Engineers, of which Pickersgill was a member, laid down the formula for calculating superheater surfaces taking the internal diameter of the element and the length inside the flue tube. Previously the method of calculation had been to take the whole external surface of the element and this should be borne in mind when comparing boiler ratios.

Following the first six engines which were built at St Rollox in 1916, were ten similar engines from N.B. Locomotive Co. Atlas Works later the same year. The numbers were 113–6, 121 and 124, 121 replacing the engine damaged at Quintinshill. The second series were Nos. 928–937. The year 1920 saw a further ten, Nos. 72–81 from St Rollox but with $20\frac{1}{2}$ in. cylinders and 180 lb. per sq. in. pressure. The class finally numbered 48 engines, Nos. 82–91 coming from Armstrong Whitworth in 1921 and Nos. 66–71 and 92–7 from N.B. Locomotive Co. Hyde Park Works in 1922. All passed into British Railways' hands and the last was scrapped in December 1962. Despite the fact that these engines had 9-in. piston valves replacing McIntosh's 8-in. valves of the 'Dunalastair IV' the Pickersgills were by no means the equal of the earlier engines.

Pickersgill's 4–6–0 engines fell into three classes, none of which can be said to have been very good. First, in 1916–17 came six outside cylinder engines, 20 × 26 in. with 6 ft. 1 in. coupled wheels. They had 24 × 5 in. and 130 × 2 in. tubes which at 15 ft. 3 in. gave 1,529 sq. ft. The firebox was 146·5 sq. ft. and the total 1,707 sq. ft. The superheater surface was 258·25 sq. ft. grate area, 25·5 sq. ft. and working pressure 175 lb. per sq. in. 56 tons 10 cwt. rested on the coupled wheels, the total engine weight being 75 tons. Not spectacular performers, these engines were given the sobriquet 'Greybacks', a greyback in Scotland being a louse, yet in 1925–6 under L.M.S. rule a further twenty were built at St Rollox, the last engines to be built in a Scottish railway works for a Scottish railway or division of one. When tested between Carlisle and Preston in 1926, No. 14630 acquitted itself well though with the heaviest coal consumption of all those tested.

In 1921 Pickersgill committed his biggest error. He built a series of four 4–6–0s, Nos. 956–9 with three cylinders. He had his own design of conjugate valve gear, the centre valve being driven off the outside Walscheart gears. It was a complex arrangement with a multiplicity of pin joints. Trouble was experienced due to fracturing of the slide bars due to stresses set up in the outside valve spindle supports. Steps were taken to remedy the trouble with little success and finally all four engines had a Stephenson Link motion fitted to drive the centre valve. This was the same pitfall Drummond had fallen into on the L. & S.W.R. The results were catastrophic and these engines never did any good. In other respects they were typical Caledonian design, 18½ × 26 in. cylinders, 6 ft. 1 in. coupled wheels and 180 lb. per sq. in. pressure. The boilers had the usual 24-element superheater and 203 × 2 in. tubes. The total heating surface was: tubes 2,200 sq. ft., firebox 170 sq. ft., total 2,370 sq. ft. The Robinson superheater provided a further 270 sq. ft. and the grate area was 28 sq. ft.

Almost equally unsatisfactory were the engines of the '191' class for the Oban road. These were saturated 4–6–0s built by N.B. Locomotive Co. at Queen's Park in 1922. The numbers were 191–8 and they had 19½-in. cylinders and 5 ft. 6 in. coupled wheels carrying 45 tons 16 cwt. of the total 62 tons 15 cwt. 2 qr. In service they proved sluggish yet they had comparatively long lap and long travel valves, the former being 1¼ in. and the latter 5½ in. with balanced slide valves instead of piston valves. The $275 \times 1\frac{3}{4}$ in. tubes had a cross sectional area to swept surface of 1 over 363 which was reasonably good, and the ratio of cross area to grate area was as high as 22 per cent. The small grate was 21·9 sq. ft. in area and typical of the grates found on Scottish engines and suitable for the coal available. The influence of the 'Rivers' is noticeable in this class.

One class of o–6–o and one tank class are ascribed to Pickersgill. The former was also a saturated engine of which St Rollox built forty-three. Nos. 300–314 came out in 1918 followed by Nos. 280, 281, 294–9, 315–324 and 670–5 in 1919. In 1920 the last four, Nos. 676–9 were built. As in McIntosh's '812' class, $18\frac{1}{2} \times 26$ in. cylinders and 5 ft. wheels were used with boilers of the same dimensions as the '812s' but with 275 tubes giving 1,332·9 sq. ft. of surface. After the 1923 Grouping some of these engines were fitted with superheaters and had the same tubing arrangements and heating surfaces as the '30' class of 1912.

Perhaps the best of Pickersgill's engines for the Caledonian were his tank engines, the 'Wemyss Bay Pugs'. These were heavy tank engines for suburban work mainly on the coast lines to Gourock and Wemyss Bay, but they were also used for banking duties at Beattock. The $19\frac{1}{2}$-in. cylinders were outside the frames, with 8-in. piston valves between, driven by Stephenson Link motion inside the frames. The intermediate valve spindles, 8 ft. 4 in. long, were cranked to pass over the leading coupled axle. All coupled axleboxes had adjustable hornblock wedges and laminated springs were provided for all axles. The boilers had 18×5 in. and $159 \times 1\frac{3}{4}$ in. tubes giving 339 sq. ft. and 1,056 sq. ft. respectively. The firebox gave 121 sq. ft. and Robinson short loop superheater elements 200 sq. ft. the grate area was 21·5 sq. ft. and the working pressure 170 lb. per sq. in. The side tanks held 1,425 gallons and the bunker tank a further 375 gallons. There was capacity for 3 tons of coal. The weight in working order was 91 tons 13 cwt. on a total wheelbase of 40 ft.

The North British locomotive department was under Reid's leadership until 1920 when he handed the reins over to the man who had been his chief draughtsman, Walter Chalmers. Chalmers was an old N.B. man who had served his time with the company and remained in various posts until made chief of the drawing office in 1904. He was responsible for following Reid's lead in the superheating of those classes which had not so far been dealt with. One of Chalmers' first actions in this respect was to remove, when rebuilding, the smokebox wing plates. In this he was not alone; Pickersgill in his later designs, Whitelegg on the G. & S.W. and Cummings on the Highland were all doing the same, thus ending a fashion which had been the vogue since the Allan 'Old Crewe' designs.

Thomas E. Heywood took charge at Inverurie on March 1, 1914. He had served an apprenticeship on the Taff Vale Railway in South Wales, joining that company in 1896. He then spent some years in Burma before returning to the T.V.R. where he became assistant locomotive superintendent.

After the last 'V' class left the Inverurie shops in March 1915 the building programme embraced only some small shunting engines for the Aberdeen docks area until the final batch of 4–4–0 were required in 1920–1. These were to be a superheated 'V' class and as such were at first classified 'VS', this was altered to 'F' before they went into service. Substantially the same as the 'V' engine but with Robinson superheaters six, Nos. 47–50, 52 and 54 were built in September and October 1920 by N.B. Locomotive Co. followed in June and September 1921 by Nos. 45 and 46 from Inverurie works.

The cylinders were the standard 18×26 in. with balanced slide valves and Stephenson Link motion. The bogie wheels were 3 ft. $9\frac{1}{2}$ in. diameter, the standard introduced by Manson in 1887, the coupled wheels were 6 ft. 1 in. and the working pressure 165 lb. per sq. in. The 4 ft. 6 in. diameter boiler, 10 ft. 6 in. long, was standard with the 'V' class and last three Manson classes. The superheater consisted of 18 elements the same as, and with the same tubing arrangements as, the previous examples. These eight engines were given names with local associations and one, No. 49, *Gordon Highlander*, has been preserved in the Glasgow Transport Museum. They were fitted with both vacuum and Westinghouse brakes. The engines weighed 48 tons 13 cwt. of which 16 tons 5 cwt. rested on the bogie which was of the swing link type.

Heywood's dock shunters were built by Manning Wardle & Co.

in 1915. There were four, two having $13\frac{1}{2}\times 20$ in. cylinders and two with 14×20 in. cylinders. They were 0–4–2Ts with 3 ft. 6 in. and 4 ft. coupled wheels respectively. The larger engines had $113\times 1\frac{3}{4}$ in. tubes at 556 sq. ft. with a 54 sq. ft. firebox and 9 sq. ft. of grate, while the smaller pair had $108\times 1\frac{3}{4}$ in. tubes at 440 sq. ft., a 47 sq. ft. firebox and a 7·75 sq. ft. grate. In each case the working pressure was 160 lb. per sq. in. and the tank capacity 450 gallons. Steam sanding was fitted and there were two Manning Wardle no. 7 injectors on the larger engines and two Gresham & Craven no. 6 combination injectors on the smaller type. Rigid wheel-bases were 6 ft. 3 in. and 5 ft. 3 in. whilst the weights were 30 tons 18 cwt. 3 qr. and 25 tons 19 cwt. 2 qr. These little engines formed G.N.S. classes 'Y' and 'X'. The first two delivered carried brass number plates 116 and 117 but were renumbered Nos. 30 and 32 in 1917. These were the larger engines. The others arrived shortly after and bore no number plates but were numbered 43 and 44 on arrival. The makers painted them in the livery adopted by Heywood, black with red and yellow lining, in place of the previously used green.

To succeed Smith at Lochgorm the Highland board chose a North British man. Christopher Cumming had served his time on the N.B. and at the time of his translation to Inverness he was district locomotive superintendent of the Fife and Northern Division of the N.B.R. His appointment was ratified at the board meeting held on October 7, 1915, and his salary was set at £550 p.a. to be reviewed after six months when in fact it was raised to £600. The same minute records the placing of orders with N.B. Locomotive Co. for three of the 'Loch' class 4–4–0 and three 'Castles'. Whilst the former were the same as the originals the latter were slightly modified having 6 ft. coupled wheels, 25·5 sq. ft. grates and 4,000 gallon tenders. The 'Lochs' were numbers 70–2 and the 'Castles' Nos. 50, 58 and 59. delivery being effected in 1916 and 1917 respectively.

The year 1917 was a busy one for the Highland. In addition to the new 'Castles' there were Cumming's first design, a 4–4–0, two of which were built by Hawthorn Leslie & Co. who also built the first four of a new type of 4–6–0. The 4–4–0s were Nos. 73 and 74 and had 20×26 in. cylinders and coupled wheels 6 ft. 3 in.

diameter. For the first time in British practice Walscheart valve gear was used outside the framing. The boilers were small for the cylinders having only 313 sq. ft. from 21 × 5 in. flues, 703 sq. ft. from 118 × 2 in. small tubes and 124 sq. ft. of firebox surface on a 22·5 sq. ft. grate which was of a drop pattern. Robinson Short Loop elements were used and contributed only 180 sq. ft. Initially the working pressure was 160 lb. per sq. in. but this was raised later to 175 lb. With 20 tons 14 cwt. 3 qr. on the bogie, 17 tons 2 cwt. 3 qr. on the drivers and 17 tons 2 cwt. on the trailers, the total weight was 88 tons 2 cwt. 3 qr. including the 3,500-gallon tender.

The passenger services thus catered for, attention was turned to the goods workings, loadings of which were very seriously increased under war-time requirements. The four 4–6–0 engines referred to above were the first of eight which had 20½ in. outside cylinders of the usual stroke, 26 in. and 5 ft. 3 in. coupled wheels. The boilers were 4 ft. 6⅝ in. diameter with 21 × 5 in. and 92 × 2 in. tubes giving 388·4 sq. ft. and 680·88 sq. ft. of surface. The Robinson superheaters added 241 sq. ft. and the fireboxes 127·3 sq. ft. The grates were 22·7 sq. ft. and the pressure 160 lb. per sq. in. This was later raised to 175 lb. as in the case of the 4–4–0s. The bogies carried 14 tons 15 cwt., leaders 13 tons 17 cwt. 1 qr., drivers 13 tons 18 cwt. 2 qr., and 13 tons 18 cwt. 1 qr. on the trailers. The engine total weight was 56 tons 19 cwt. and the tender, which carried 3,000 gallons of water and 5 tons of coal, weighed 35 tons 13 cwt. 1 qr. Originally fitted with screw reverse, the first four engines had this changed for lever reverse about the time, in 1919, when the second four were delivered. The numbers were 75–82.

Cumming's last design was the 'Clan' class 4–6–0. Again eight engines were delivered in two batches, four in 1919 and four in 1921 and the builders were Hawthorn Leslie & Co. Intended for fast trains on the main line to the south they had 6-ft. coupled wheels, 21 × 26 in. cylinders with 10-in. piston valves and outside Walscheart valve gear. The bogies reverted to the old practice of having the centre pin in advance of the centre line of the bogie wheelbase though this time only ½ in. instead of the more usual 1 in. The coupled wheelbase was 14 ft. and the driving tyres were provided with thin flanges on the lines of the later dictum of the

A.R.L.E. Laminated springs were fitted to all axles except the drivers which had spiral springs. The boilers were adequate. Containing 21×5 in. tubes equal to 409 sq. ft. and 118×2 in. tubes giving 919 sq. ft., they were pressed to 175 lb. per sq. in. The fireboxes had a surface of 139 sq. ft. and the Robinson superheaters 256 sq. ft. and the grate area was 25·5 sq. ft. These engines had steel fireboxes which gave a lot of trouble, at one time in 1922 there were five of them out of service at Aviemore with severely burned fireboxes. Later these engines were fitted with new copper boxes. The weights were 62 tons 4 cwt. 3 qr., of which 45 tons 9 cwt. 3 qr. was available for adhesion. The first batch comprised Nos. 49 and 51–3 and the second Nos. 54–7. Four of the class had Westinghouse brakes. The goods engines built earlier had cost only £4,920 each but these, with the mounting costs during the War, were £6,957 each. In 1921 No. 53 was experimentally fitted with 'Scarab' equipment for burning oil fuel with which apparatus a train of 241 tons was taken from Aviemore to Inverness via Forres on a fuel consumption of 31 lb. oil per mile. The calorific value of the oil was 18,900 B.T.U. per lb. and the results compare favourably with the normal 45 lb. per mile of coal. This test was carried out before the coal strike of 1921, possibly in anticipation of such an emergency, not because of it.

Cumming retired on health grounds in 1922 and was succeeded by David Chalmers Urie who held office until the Grouping at the end of the year. Urie then became mechanical engineer of the Northern Division of the L.M.S.R. During his short term of office he carried out some useful rebuilding of the 'Big Ben' class into which he put 19 element superheaters. The boiler data then became: heating surface, tubes 746·9 sq. ft. and 270·8 sq. ft., firebox 132 sq. ft. and superheater 168·4 sq. ft. They kept the same grates but the working pressure was lowered to 175 lb. per sq. in.

It was during Urie's time at Inverness that the 'Rivers' returned to the line for which they had originally been intended, a measure of bridge strengthening having been carried out in the interim. It is, however, interesting to note that the hammer blow of these engines was very favourable compared with that of the 'Castles' or the 'Clans', in fact the combined load on the rail at 6 r.p.s., about 78 m.p.h., was actually 1 ton less than that of the 'Clan' class, while the hammer blow for the whole engine and per axle was considerably less than both classes mentioned. So at last Smith's foresight in design was vindicated.

In the heavily forested areas through which the Highland ran, fires caused by sparks from locomotives were a severe hazard. In 1901 Drummond had brought up the question of fitting engines with his brother's spark arresters, claiming a saving in fuel as well as freedom from throwing sparks. The apparatus was made by the Glasgow Railway Engineering Company, a Drummond firm run by Dugald's sons George and Walter. At first there was a reluctance to pay the royalty of £5 per engine. However in 1902 it was agreed to fit a number of engines. In 1907 the Locomotive Committee recommended the discontinuation of the appliance and the allocation of a sum of £500 per annum to be set aside to meet liabilities arising from claims for damage from line-side fires. The Glasgow firm offered retention of the fitting on 100 engines at £2 per engine per annum and this was confirmed. The difference between the old and new royalties was to be put into a special insurance fund and finally in August 1913 the agreement with the manufacturers of the apparatus was terminated and an annual sum of £100 paid into the Engine Sparks etc. Insurance Fund. In 1918 this sum was increased to £300.

On August 6, 1918, the G. & S.W. directors minuted the appointment of Robert H. Whitelegg as successor to Drummond, as from August 1st. Born at Garston, Lancashire, in October 1872 and educated in London he had joined the London Tilbury & Southend Railway as an apprentice in 1887. After having had some drawing office experience he was made inspector of new rolling stock and materials and located at Birmingham in 1891. The following year he was at Naysmith Wilson's works, Manchester, supervising the construction of engines being built for the L.T. & S. and he followed this with a short period in Spain for the same firm. In 1905 Robert became assistant to his father Thomas at Plaistow, with special responsibility for the reorganization of the locomotive running department.

When Whitelegg assumed the mantle of responsibility at Kilmarnock the war was nearing its end and the condition of the locomotive stock such that a great deal of repair work was urgently required; engines had to be kept at work for long periods after they were normally due for shopping and many boilers were worn out and working at reduced pressures. One of his first essays was to produce a series of new boilers: the X1 boiler suitable for the '8', '240' and '336' classes, the X2 for the '17' and '361' classes, and the X3 for the '119', '153', '22', '160', and '306' classes. In the X1 boiler the heating surfaces were increased from the original amounts by around $12\frac{1}{2}$ per cent and the working pressure raised to 170 lb. per sq. in.

At the same time Whitelegg modified the valve gear, removing the rocker shafts and opening the eccentric rods. The valves were now driven by an off-set spindle on an arm about 1 ft. 3 in. from the intermediate valve spindle. Instead of improving the distribution of steam and reducing trouble from the greater number of pin joints, a large number of failures occurred in which the off-set arms broke. Moreover it has often been said that the valve setting under Whitelegg's management was far inferior to that which had obtained previously.

Having had considerable experience on the L.T. & S. of mass movement of the populace over comparatively short distance Whitelegg found a very similar feature in the commuter traffic between Glasgow and the Ayrshire coast towns and to Kilmarnock. To deal with this still growing problem he designed a series of large 4-6-4 tank locomotives based, inevitably on the same type on the L.T. & S.R. When these massive engines arrived from N.B. Locomotive Co's works in 1922 they were received with acclaim by officials and footplate staff of a company hitherto more inclined to look askance at tank engines. They were indeed beautiful machines and quickly showed their capabilities.

Two outside cylinders 20×26 in. drove on to the middle coupled axle. Walschearts valve gear operated the 10 in. diameter inside admission piston valves which were the patent of Allen & Simmonds, Reading. The coupled wheel bearings were of generous size, $8\frac{1}{2}$ in. diameter and $10\frac{1}{2}$ in. long, and laminated springs were used. The bogies had coil springs and were given $5\frac{1}{2}$ in. side play. Whitelegg used the swing ling type of bogie introduced by Manson in the interests of standardisation though he would have preferred the Adams type. In his own words the decision to use two instead of three cylinders of more moderate size was dictated by cost rather than weight, and of course the simplicity got from the lesser number of sets of motion. The boilers were 5 ft. $6\frac{3}{16}$ in.

diameter and 14 ft. 11 in. long. The heating surface from $21 \times 5\frac{1}{4}$ in and 141×2 in. tubes was 1,574 sq. ft., and of the firebox 156 sq. ft. The grate area was 30 sq. ft. and the pressure 180 lb. per sq. in. Robinson Short Loop elements gave a superheater surface of 255 sq. ft. and Robinson header discharge valve and draught retarders were fitted. The water tanks held 2,400 gallons and the coal bunker $3\frac{1}{2}$ tons.

Part of the splendour of these engines lay not so much in their size as in their appearance which was enhanced by the blue steel planished lagging plates which blended well with the green and maroon of the company's livery. On the footplate controls were duplicated so that the engines could be driven from either side with ease, there being regulator handle, brake valve, whistle and sanders on each side of the cab. The weight of the engines was 99 tons 1 cwt. 2 qr., of which 21 tons 6 cwt. rested on the leading bogie and 23 tons 15 cwt. 2 qr. on the trailing bogie. The coupled axles carried 18 tons each on a wheelbase of 13 ft. 2 in. in a total of 39 ft., the overall length being 47 ft. $4\frac{1}{4}$ in.

Manson's four-cylinder No. 11 was successfully rebuilt by Whitelegg with several modifications including re-cylindering with four 14 in. diameter cylinders retaining the original strokes of 24 in. outside and 26 in. inside. By introducing cross ports it was possible to utilize one piston valve for each pair of cylinders eliminating the need for rocking shafts. Again the piston valves were Allen & Simmonds patent. These valves were arranged so that the rings were subjected to pressure on both sides, i.e. effective pressure on the front and back pressure on the underside preventing excessive pressure of the rings on the valve liners. The rings each comprised a steam ring and an exhaust ring, each was narrower on the contact face than the underside which was exposed to steam pressure. Both were held against the liners by springs and the sections of the valve head were also spring loaded.

As rebuilt, No. 11 had a 22-element Robinson superheater and $187 \times 1\frac{3}{4}$ in. tubes. The total tube surface was 1,443 sq. ft. and that of the firebox 148 sq. ft., a considerable improvement on the earlier boiler. The grate was 27·6 sq. ft. in area and the pressure 180 lb. per sq. in. From this it will be seen that the practice of dropping the pressure on superheating had at last been discontinued. The rebuilt engine weighed 61 tons 9 cwt. which, with the 37 tons 7 cwt. of the tender, made a total of 98 tons 16 cwt. The tender carried 3,260 gallons and 5 tons of water and coal. It carried the name *Lord Glenarthur*.

Whitelegg would have liked to build more four-cylinder engines, to have developed his scheme for reboilering existing classes and to have incorporated a 'Baltic' boiler in a Manson 4-6-0. This would have entailed redesigning the frames and strengthening them throughout. Besides these he would have tried to enlarge his big tank engines and put a six-wheel bogie under the trailing end to allow for greater coal and water capacity. All these ideas were however nullified by the passing of the Railway Act of 1921 by which the railways of Britain were 'grouped'. The Glasgow & South Western, Caledonian and Highland found themselves in the new London Midland & Scottish Railway and the Great North of Scotland and North British in the London & North Eastern Railway. With this event, rendered necessary by a matter of sheer economics, all new design and construction in the works of the Scottish companies for their own purposes ceased and these works became repair and maintenance works only.

Throughout the ninety-two years during which the Scottish railways had existed there had been many and varied designs of locomotive, some very good, some not so good, and many contributions had been made to the science of engineering as applied to the steam locomotive.

Index

Aberdeen, race to, 31, 145, 168, 188
Aberdeen Rly, 41, 48, 84
Accidents, 41, 90, 98, 120, 159, 203, 207
Allan, A. 55–6
Andrew Barclay & Co., 38, 66, 150–1
Arbroath & Forfar Rly, 84
Arbroath Works, 60, 84
Ardrossan Canal, 20
Armstrong Whitworth & Co., 224
Ayrshire & Wigtownshire Rly, 122

Ballochney Rly, 19
Banff, Portsoy & Strathisla Rly, see
    Banffshire Rly
Banffshire Rly, 62
Barclay, W., 57–8
Belgian State Rlys, 191
Berwickshire Rly, 25
Bessemer, H., 56
Beyer, C., 71
Beyer, Peacock & Co., 59, 75, 97
Black, Hawthorn & Co., 96
Border Counties Rly, 25
Border Union Junction accident, 159
Border Union Rly, 25
Bridge of Weir Rly, 23, 69
Brittain, G., 106–10
Brown, W. Steel, 74
Burntisland Works, 78
Bury, Curtis & Kennedy, 39, 43

Cairn Valley Light Rly, 163, 165–6
Caithness Rly, 27
Caledonian Dunalastairs, 207
Caledonian & Dunbartonshire Rly, 77
Caledonian Rly, 20–2, 30–3, 44–6, 51,
    64–6, 84–7, 106–11, 127–35, 146,
    183–209, 223–6
Callander & Oban Rly, see Caledonian
    Rly
Callander & Oban Railway, 108
Canals, 18, 19, 20, 23, 30, 36–7
Carmichael & Co., 37
Castle Douglas & Dumfries Rly, 23, 69
Chalmers, W., 226–7
Chemin de fer l'Est, 180
City of Glasgow Union Rly, 23, 26, 100
Clark, D. K., 58–9, 61, 73–4
Clyde Locomotive Co., 39, 122, 139
Conner, B., 64–6, 83–7
Cook Street Works, 33, 51

Costs, Operating & Repair, 146
Cowan, W., 60–1, 105–6
Cowlairs Works, 33, 40, 78–9, 82, 89–92,
    95–6, 111, 137–8, 168–9, 171, 175–8,
    206, 215–17
Cumming, C., 228–30

Design
  Balancing, 85, 158
  Bogies, 60–1, 93, 104, 107, 116, 120–1,
    125, 232
  Boiler Barrels, 42–3, 111, 200, 232
  Boiler Tubes, 174, 183, 198, 200
  Brakes, 20, 31, 40, 46, 86, 92, 94, 98, 99,
    104, 105, 120, 122, 123, 124, 134, 136,
    137, 140, 144, 152, 168, 174, 177, 179,
    180, 182, 187, 192, 195, 212, 221, 227,
    230
  Cabs, 53–4, 82, 113
  Chimneys, 104, 114
  Compounding, 138–9, 171–2
  Condensing, 184, 199
  Connecting & Coupling Rods, 67, 112,
    192, 196, 211
  Crampton type, 47
  Cross Water Tubes, 156–7, 179
  Crown Stays, 43, 111–12, 194, 202
  Cylinders & Valves, 44, 65, 125, 129,
    138–9, 141, 158, 164, 195, 232–3
  Domes, 43, 70–1, 112
  Drop Grate, 111, 221
  Eccentric Crank Pins, 59, 81
  Feed Pumps, 45, 211
  Feed Water Heating, 179, 211, 222–3
  Firebars, 45, 212
  Fireboxes, 42, 56, 156, 160, 161, 230
  Frames, Bar, 42, 67; Double, 71; Plate,
    111, 160–1, 170, 196; Sandwich, 42
  Hand Brakes, 40, 46, 66
  Lagging, 44
  Le Chatelier Counter Pressure Brake,
    140
  Lubrication, 130, 153, 174, 212
  Oil Fuel, 193, 195, 230
  Old Crewe type, 44, 55, 84, 102
  Regulator Valve, 45, 194
  Reversing Shaft, 120, 131
  Safety Valves, 45, 72, 92, 112, 123, 140,
    151
  Smokeboxes, 114, 119, 196

Smoke Prevention, 61–2, 73, 123, 149, 163
Spark Arresters, 114, 231
Springs, 46, 53, 67, 85, 104, 110, 115, 136, 137, 168, 193
Steam Brakes, 92, 94, 120, 137, 177
Steam Pipes, 44–5, 100, 139
Steam Reverser, 86, 113, 180, 221
Steel-McInnes Brake, 31, 86
Superheating, General, 205–6; Cal., 205, 206–9, 224–5; G.N.S., 218–19, 227; G. & S.W., 204, 213, 232–3; High., 206, 228–30; N.B., 174–5, 206, 215–17
Tablet Apparatus, 29, 126–7
Tenders, 46, 53, 87, 115, 126, 140, 155–6, 162, 190
Tyre Fastenings, 129, 192
Vacuum Brake, 31, 98, 104, 137, 140, 144, 179, 180, 221, 227
Valve Gears, 46, 54, 55–6, 131, 149, 155, 158, 161, 212, 225, 232
Westinghouse Brake, 31, 86, 94, 98, 99, 122, 123, 124, 134, 136, 137, 144, 152, 174, 177, 179, 180, 182, 195, 212, 221, 227, 230
Deeside Rly, 63–4
Dingwall & Skye Rly, 27, 103, 105
Dodds, Isaac, 35
Drummond, D., 31, 88, 91–7, 127–35
Drummond, P., 177–83, 210–14
Dübs & Co., 39, 70, 80, 85, 86, 95, 96, 98, 103, 108, 125, 131, 143, 154, 156, 180
Dunbar, A.G., 131
Dundee & Newtyle Rly, 22, 37
Dundee, Perth & Aberdeen Junction Rly, 22, 47, 49
Dundee & Perth Rly, see D.P. & A.J.R.

Eastern Counties Rly, 51, 61, 62, 63
Edinburgh Exhibition 1886, 131, 136, 139
Edinburgh & Glasgow Rly, 33, 39, 40, 71–4
Edinburgh & Hawick Rly, 25
Edinburgh & Northern Rly, see E.P. & D.R.
Edinburgh, Perth & Dundee Rly, 24, 78
Edinburgh, Race to 1888, 31, 145
Edington & Co., 39, 44
Engineer, The, 139, 203
Engineering, 146, 158, 203
Engineers, *see* under uames of indiviudal engineers
Engineers & Shipbuilders in Scotland, Institute of, 34
England & Co., 48–9

Exhibitions, 65, 131, 136, 139

Fairbairn & Co., 58
Fife & Kinross Rly, 24
Findhorn Rly, 57
Forth Bridge, 25, 136
Forth & Clyde Canal, 18, 19, 30
Forth & Clyde Junction Rly, 26, 90–1

Girvan & Portpatrick Junction Rly, 23, 121–2
Glasgow, Bothwell, Hamilton & Coatbridge Rly, 96
Glasgow, Dumbarton & Helensburgh Rly, 77
Glasgow, Dumfries & Carlisle Rly, 21–2
Glasgow & Garnkirk Rly, 21, 36
Glasgow, Paisley & Greenock Rly, 20, 38, 50, 133
Glasgow, Paisley, Kilmarnock & Ayr Rly, 20–2, 38–9
Glasgow & South Western Rly, 22–3, 31–2, 50–4, 69, 97–101, 116–22, 146, 153–66, 205–6, 210–15, 223, 231–4
Glasgow & South Western Rail Motors, 164–6
Gogar accident, 41
Grand Junction Rly, 20, 55
Greenock & Ayrshire Rly, 23
Greenock Works, 51, 110
Great Central Rly, 207
Great Eastern Rly, 76, 149
Great Northern Rly, 41, 74, 75
Great North of Scotland Rly, 28–9, 58–64, 105–6, 122–7, 147–53, 218–19, 227–8
Great North of Scotland Rail Motors, 150–2
Great Western Rly, 217

Hawthorns of Leith, 33, 39, 43, 57, 74, 82, 103
Hawthorn & Co., R. & W., 52, 53, 54
Hawthorn Leslie & Co., 220, 228, 229
Heywood, T.E., 227–8
Highland Rly, 27, 29, 33, 81–3, 101–5, 139–45, 177–83, 219–22, 228–31
Holmes, M., 135–9, 167–71
Holytown Bridge Works, 183
Hurst, W., 69–71, 79

Institution of Engineers & Shipbuilders in Scotland, 34
Ironworks, 17–18
Inverness & Aberdeen Junction Rly, 27, 57
Inverness & Nairn Rly, 27, 56
Inverness & Perth Junction Rly, 27, 57

Inverness Works, see Lochgorm Works
Inverurie Works, 150, 227

Johnson, J., 147–9
Johnson, S. W., 74–6
Jones, D., 101–5, 139–45, 147
Jones & Potts, 41, 77

Kilmarnock & Troon Rly, 19, 37
Kilmarnock Works, 51–4, 67, 97–9, 101,
   116–19, 122–3, 153–5, 163–4, 212,
   231–2
Kinmond, Hutton & Steel, 39, 44
Kinross-shire Rly, 24
Kipps Works, 36
Kirkcudbright Rly, 23, 69
Kitson & Co., 123–5, 144
Kittybrewster Works, 59, 105, 124, 126,
   147

Lambie, J., 183–6
Lancaster & Carlisle Rly, 20
Lancaster & Preston Rly, 20
Leeds & Selby Rly, 76
Leven & East of Fife Rly, 24, 96
Liverpool & Manchester Rly, 20, 42
Lochgorm Works, 80, 82, 101, 103, 105,
   139, 143, 145, 177, 181, 183, 219, 228,
   230
Locke, Joseph, 20, 38
Locomotives
   Banffshire, 0–4–2, 62; 64; 0–4–2T, 62;
   Caledonian, *Old Crewe* type, 44–6;
      Conner, 2–2–2, 64–6; 2–4–0, 86–7;
      Tank engines, 87; Brittain, 4–4–0,
      107–8; Crane engine, 108, 0–4–2,
      108; 2–4–0T, 108; 4–4–0, 'Oban
      Bogie', 109–110; 2–2–2T Rebuild,
      110; Drummond, 2–4–0 Rebuild,
      128; 'Jumbo' 0–6–0, 128–130; 4–4–0,
      130–1; 4–2–2, 132–3; 4–4–0 'Green-
      ock Bogie', 133; 0–4–4T, 133–4;
      Dock Pugs, 134; 0–6–0T, 134; 0–4–
      2T, 134; Lambie, 4–4–0T, 183; 0–4–
      4T, 184; 0–6–0T, 184; 0–6–0ST,
      185; 4–4–0, 185–6; McIntosh, 4–4–0
      'Dunalastair' I, 31, 188–9; 'Dunala-
      stair' II, 189–90; 'Dunalastair' III,
      190–1; 'Dunalastair' IV, 191–2; 0–6–
      0, 194–6; 0–8–0, 196–7; 0–8–0T, 197–
      8; 0–4–4T, 198–9; 0–6–0T, 199; 4–
      6–0, Oban Bogie, 199–200; 49 cl,
      200–1; Cardean, 242–3; 908 & 918 cl,
      203–4; Superheater, 4–4–0, 205–8;
      4–6–0, 179 cl, 208–9; 0–6–0, 209;
      Pickersgill, 4–4–0, 224; 4–6–0, 60 cl,
   225; 956 cl, 225; 'Oban Bogie', 191 cl,
   225–6; 0–6–0, 227; 4–6–2T, 227;
Callander & Oban, 2–4–2T, 108–9;
Caledonian & Dumbartonshire, 2–2–2T,
   77
Deeside, 0–4–2T, 63; 0–4–2, 63–4
Dundee & Newtyle, 0–2–4, 0–4–2, 37
Edinburgh & Glasgow, 'Bury' type, 39–
   40; Paton 0–6–0T, 40; England, 2–2–
   2, 48–9; 2–4–0, 72; 0–4–2, 72; 0–4–0,
   73; Brown, 2–4–0, 74; Johnson, 2–4–
   0, 75–9; 0–4–2, 75; 0–4–0, 75–6
Edinburgh, Perth & Dundee, Nicholson, 0–6–0, 83
Forth & Clyde Junction, 2–4–0, 91;
Girvan & Portpatrick Junction, Wheatley, 0–6–0, 122
Glasgow, Bothwell, Hamilton & Coatbridge, 0–6–0T, 96
Glasgow Dumbarton & Helensburgh,
   2–2–2, 77
Glasgow & Garnkirk, 2–2–0, 0–4–0,
   35–6
Glasgow, Paisley, Kilmarnock & Ayr,
   Miller, 2–2–2, 39
Glasgow & South Western, P. Stirling,
   2–2–2, 95 cl, 52; 0–4–0, 99 cl 52;
   0–6–0, 103 cl, 53; 2–4–0 *Galloway*,
   52; 0–4–2, 22 cl, 53; 2–2–2, 40 cl, 53;
   0–4–2, 32 cl, 53; 131 cl, 54; 141 cl,
   54; 2–2–2, 45 cl, 54; 0–6–0, 58 cl, 54;
   0–4–0, 52 cl, 54; J. Stirling, 2–4–0,
   71 cl, 97; 8 cl, 97; 0–4–2, 208 cl, 98;
   187 cl, 98; 221 cl, 98; 0–4–0, 65 cl,
   98–9; 0–4–0T, 14 cl, 98; 4–4–0, 6 cl,
   99–100; 0–6–0, 13 cl, 100; 0–4–0T,
   1 cl, 101; Smellie, 2–4–0, 187 cl, 116–
   17; 0–6–0, 22 cl, 117–18; 4–4–0,
   119 cl, 118–19; 153 cl, 119–20; Manson, 0–6–0, 306 cl, 153; 160 cl, 153–4;
   361 cl, 154; 17 cl, 154; 4–4–0, 8 cl,
   154–6; 336 cl, 156–7; No. 11, 157–9;
   240 cl, 159–60; 18 cl, 160; 4–6–0,
   381 cl, 160–2; 0–4–4T, 326 cl, 162–3;
   266 cl, 163; 0–6–0T, 14 cl, 164; 0–4–
   0T, 272 cl, 164; Rail Motor, 164–6;
   Superheater, 4–6–0, 128 cl, 210;
   Drummond, 0–6–0, 279 cl, 211; 4–4–
   0, 131 cl, 212; 137 cl, 212–13; 2–6–0,
   213; 0–6–2T, 213; 0–6–0T, 214;
   Whitelegg, 4–6–4T, 32, 232–3; 4–4–0
   Rebuild, 233–4
Great North of Scotland, Clark, 2–4–0,
   57; Ruthven, 0–4–0WT, 58; Cowan,
   2–4–0, 59; 4–4–0, cl H & K, 59–61;

cl L, 105–6; cl M, 106; Manson, 4–4–0, cl A, 123; cl G, 124; 0–6–0T, cl D & E, 124; 4–4–0, cl N, 124–5; cl O, 125; cl P & Q, 125–6; Johnson, 0–4–4T, cl R, 148; 4–4–0, cl S, 148–9; Pickersgill, 4–4–0, cl T, 149; cl V, 150; Rail Motor 150–2; Superheater, 4–4–0, 218–19; Heywood, 4–4–0, cl F 227; 0–4–2T, 227–8

Highland, Stroudley, Rebuilds, 81, 90; 0–6–0T, 83, 120; Jones, 2–4–0, Rebuilds, 103; 4–4–0, cl F, 103–4; 2–4–0T, 105; 4–4–0, 'Skye Bogie' 105; 'Bruce' cl, 139–40; 'Strath' cl, 140–1; 'Loch' cl, 141–2; 0–4–4T, 142; 4–4–0T, 142–3; 4–6–0 'Big Goods', 143–4; 2–4–0T, 145; Drummond, 4–4–0, 'Wee Ben', 178–9; 0–6–0, 'Barney', 179; 4–6–0, 'Castle' cl, 180–1, 219; 0–6–0T, 181; 0–6–4T, 182; 4–4–0, 'Big Ben', 182, 222, 230; Smith, 4–6–0, 'River' cl, 220–2; Cumming, 4–4–0, 228–9; 4–6–0, Goods, 229; 'Clan', 230–1

Inverness & Nairn, Barclay, 2–2–2, 2–4–0, 57

Inverness & Perth Junction, 2–2–2, 2–4–0, 57; 0–4–0T, 57

Kilmarnock & Troon, 37

Monkland, 0–4–2, 78; 0–6–0, 83

Monkland & Kirkintilloch, Dodds, 0–4–0, 35

Morayshire, Samuel, 2–2–0T, 63; 2–4–0, 63

North British, Hurst, 2–4–0, 70; 0–6–0, 70; 2–2–2WT, 70–1; 0–4–2T, 71; 0–4–0, 78–9; Wheatley, 0–4–0, 79; 0–6–0, 80, 90; 4–4–0, 88–9; 2–4–0; 89; 0–4–0T, 90; 0–6–0ST, 90; Drummond, 0–6–0T, 91–2; 0–6–0, 18 in, 92; 2–2–2, 92–3; 4–4–0, 93–4; 0–4–2T, 95; 4–4–0T, No. 494, 95; No. 72, 95; 0–6–0, 17 in. 95–6; Rebuilds, 96–7; Holmes, 4–4–0, 574 cl, 136; 592 cl, 136; 0–6–0, 17 in. 137; 18 in. 137–8; 0–4–0T, 138; 0–4–0ST, 138; Compound, No. 224, 138–9; 4–4–0 West Highland Bogie, 167–8; '633' cl, 168; '729' cl, 169; '317' cl, 169–70; 0–6–0T, 169; Reid, 4–4–2, 32, 172–3; 4–4–0, 173–5; 0–6–0, 175–6; 0–4–4T, 176; 0–6–2T, 176–7; 4–4–2T, 177; Superheater 4–4–0, 'Scott', 215–16; 'Glen', 215–16; 0–6–0, 217; 4–4–2T, 218

Paisley & Renfrew, 2–2–0, 2–2–2T, 38

Ross-shire, 2–2–2, 2–4–0, 88

Scottish Central, Allan, 55

Scottish Midland Junction, Yarrow, 0–4–2, 84; 2–4–0, 84–5

Locomotives (Named)
Aldourie, 57, 103; Atalanta, 77; Ballindalloch Castle, 206; Balnain, 83; Ben Nevis (Ben-y-Gloe), 178; Berwick, 92; Borderer, 173; Breadalbane, 103; Bruce, 139; Cardean, 202; Colville, 144; Dunalastair, 89; Earl of Airlie, 37; Eglinton, 131; Frew, 36; Galloway, 52; Garnkirk, 36; Gartgill, 36; George Stephenson, 36; Glasgow (G & G), 36; Glasgow (NB), 92; Glen Douglas, 216; Gordon Castle, 145; Gordon Highlander, 227; Hercules, 40, 177; Jenny, 36; John Bull, 38; Kinmundy, 125; Kinnaird, 48; Kirkintilloch, 35; Little Scotland, 48; Lochgorm, 83; Lord Glenarthur, 234; Lord Wharncliffe, 37; Monkland, 35; Napier, 41; Raigmore, 57, 81, 83, 102, 117; Redgauntlet, 173; River Ness, 222; St. Martins, 83; St. Rollox, 36; Samson, 40; Sirius, 48; Sir James Thompson, 202; Thomas Adam, 125; Trotter, 37; Victoria, 36

London & Birmingham Rly, 42

London Brighton & South Coast Rly, 88, 91, 92, 95, 178

London Exhibition, 1861, 65

London Midland & Scottish Rly, 181, 195, 230, 234

London & North Eastern Rly, 79, 89, 170, 217, 234

London & North Western Rly, 77, 146, 108–9, 188, 203, 205

London & South Western Rly, 178, 214, 280

London, Tilbury & Southend Rly, 231, 232

Longsdon, Alfred, 56

Maidens & Dunure Light Rly, 163

McIntosh, J. F., 31, 187–209

Manchester, Sheffield & Lincolnshire Rly, 75, 76

Manning, Wardle & Co., 227–8

Manson, J., 32, 122–7; 153–66, 205, 210

Manuel Accident, 90

Maryport & Carlisle Rly, 67–8, 123

Maybole & Girvan Rly, 23, 69

Metropolitan Rly, 60

Midland Rly, 89, 93, 94, 146, 147, 148
Miller, J., 39, 43
Monkland Canal, 18, 19, 36–7
Monkland & Kirkintilloch Rly, 19, 35
Monkland Rlys, 76, 77–8
Monterau & Troyes Rly, 205
Morayshire Rly, 62, 63
Murdoch & Aitken, 35

Namur & Liege Rly, 47
Nasmyth, Wilson & Co., 231
Neilson & Co., 63, 66, 80, 85, 86, 92, 98, 105, 106, 107, 122, 130, 137, 148, 169
Neilson & Mitchell, 39, 64
Neilson, Reid & Co., 39, 154, 191
Newcastle & Berwick Rly, 24
Newcastle & Carlisle Rly, 25
Newlands, Alex., 220
Nicholson, R., 78
Nock, O. S., 207
North British, Arbroath & Montrose Rly, 25
North British Locomotive Co., 39, 154, 160, 175, 176, 182, 210, 212, 213, 214, 217, 219, 224–5, 227, 232
North British Rly, 23, 24, 26, 30–2, 69–70, 78, 88–94, 96, 135–9, 167–77, 215–18, 226–7
North Eastern Rly, 25, 146, 171, 173, 217–18, 219
North London Rly, 61
Northumberland Central Rly, 25

Operating Costs, 146

Paisley & Renfrew Rly, 38
Perth & Dunkeld Rly, 27, 80
Perth Works, 84
Paton, W., 40, 71–3
Peto, Brassey & Betts, 91
Petre, Hon. E. G., 50, 69
Pickersgill, W., 149–53, 219, 223–6
Plateways, 19
Pollok & Govan Rly, 49
Port Carlisle Rly, 24
Portpatrick Rly, 22, 49

Quintinshill accident, 203, 207

Race to the north, 1888, 1895, 31, 145, 168, 188
Rail Motors, 150–2, 164–6
Reid, James, 34
Reid, W. P., 32, 171–7, 215–18
Repair Costs, 146
Robinson, P., 50, 53
Ross-shire Rly, 27, 82

Ruthven, J. F., 50–60

St. Andrews Rly, 24
St. Combs Light Rly, 124
St. Enoch accident, 98
St. Margaret's Works, 48, 69, 70, 79
St. Rollox Works, 51, 86, 106, 108, 109, 121, 128, 130, 133, 134, 135, 183, 189, 190, 192, 195, 196, 197, 200, 205, 208, 223, 224
Samuel, J., 63
Scottish Central Rly, 21, 27, 55, 81, 82, 83, 84
Scottish Midland Junction Rly, 21, 81, 84
Scottish North Eastern Rly, 41, 81, 84, 187
Scott, Sinclair & Co., 39
Sharp, Roberts & Co., 37, 39
Sharp, Stewart & Co., 39, 57, 125, 143, 154
Silloth Bay Rly, 25
Sinclair, Robt., 50–1
Slamannan Rly, 17
Smellie, Hugh, 31, 32, 116–21, 198
Smith, D. L., 118, 163
Smith, F. G., 219–22
Snowploughs, 82–3, 137–8
South Eastern & Chatham Committee, 150
South Eastern Rly, 62, 101
Southside & Suburban Rly, 26
Stark & Fulton, 39, 44
Steel, 56, 230
Stephenson & Co., Robt., 38, 60, 125, 173
Stirling & Co., 37
Stirling & Dunfermline Rly, 77
Stirling, J., 32, 97–101
Stirling, P., 51–4
Strathspey Rly, 28
Stroudley, W., 71, 76, 80–3, 104–5
Superheating, 205–34
Sutherland Rly, 27, 145

Taff Vale Rly, 227
Tales of the G. & S.W.R., 118
Tay Bridge, 25, 89
Tayleur & Co., see Vulcan Foundry
Telford, Thomas, 18
Tests
  Cardean, CR, 203
  Drummond, 4–4–0, CR, 185
  Dunalastair II, CR, 190
  Feed Water, CR, 222
  Feed Water, Highland, 222
  Feed Water, G. & S.W., 223

Lambie, 4-4-0, CR, 185-6
McIntosh Superheater, 4-4-0, CR, 206-7
Little Scotland E. & G., 48
Highland/NB, 174, 181
Highland Oil Fired, 230
NB/L. & N.W., 173
NB/N.E./G.W. Goods, 217
Thomas, John, 108
Thornton, Robt., 41-2, 50
Tulk & Ley, 47-8

Union Canal, 18, 19
Urie, D. C., 182, 230

Vulcan Foundry, 41, 43-4
Wansbeck Valley Rly, 25
Wardale, J. D., 60
Watt, James, 17
West Highland Rly, 167
Wheatley, Thomas, 76-80, 88-91, 122
Wheatley, W. T., 122
Whitelegg, Robt., 32, 231-4
Wilson, E. B. & Co., 48
Wishaw & Coltness Rly, 21, 49, 183

Yarrow, Thomas, 84-5
York Newcastle & Berwick Rly, 41
Yorkshire Engine Co., 177